Bioinformatics: Sequence, structure, and databanks

'Atomics is a very intricate theorem and can be worked out with algebra but you would want to take it by degrees because you might spend the whole night proving a bit of it with rulers and cosines and similar other instruments and then at the wind-up not believe what you had proved at all.

'Now take a sheep', the Sergeant said. 'What is a sheep only millions of little bits of sheepness whirling around and doing intricate convolutions inside the sheep? What else is it but that?'

(from *The Third Policeman*, Flann O'Brien)

The Practical Approach Series

Related **Practical Approach** Series Titles

Protein-Ligand Interactions: structure and spectroscopy*

Protein-Ligand Interactions: hydrodynamic and calorimetry*

DNA Protein Interactions

RNA Protein Interactions

Protein Structure 2/e

DNA and Protein Sequence Analysis

Protein Structure Prediction

Antibody Engineering

Protein Engineering

* indicates a forthcoming title

Please see the **Practical Approach** series website at
http://www.oup.co.uk/pas
for full contents lists of all Practical Approach titles.

Bioinformatics: Sequence, structure, and databanks

A Practical Approach

Edited by

D. Higgins

Department of Biochemistry,
University College, Cork,
Ireland

and

W. Taylor

Division of Mathematical Biology,
National Institute for Medical Research,
The Ridgeway, Mill Hill,
London NW7 1AA, UK

OXFORD

UNIVERSITY PRESS

OXFORD
UNIVERSITY PRESS

Great Clarendon Street, Oxford OX2 6DP

Oxford University Press is a department of the University of Oxford.
It furthers the University's objective of excellence in research,
scholarship, and education by publishing worldwide in

Oxford New York

Athens Auckland Bangkok Bogotá Buenos Aires Calcutta Cape Town
Chennai Dar es Salaam Delhi Florence Hong Kong Istanbul Karachi
Kuala Lumpur Madrid Melbourne Mexico City Mumbai Nairobi Paris
São Paulo Singapore Taipei Tokyo Toronto Warsaw

with associated companies in Berlin Ibadan

Published in the United States by Oxford University Press Inc., New York

First published 2000

A catalogue record for this title is available from the British Library

Library of Congress Cataloguing in Publication Data (Data available)

1 3 5 7 9 10 8 6 4 2

ISBN 0 19 963791 1 (Hbk.)
ISBN 0 19 963790 3 (Pbk.)

Typeset in Swift by Footnote Graphics, Warminster, Wilts
Printed in Great Britain on acid-free paper
by The Bath Press, Avon

Preface

Bioinformatics an emerging field

In the early eighties, the word 'bioinformatics' was not widely used and what we now know as bioinformatics, was carried out as something of a cottage industry. Groups of researchers who otherwise worked on protein structures or molecular evolution or who were heavily involved in DNA sequencing were forced, through necessity, to devote some effort to computational aspects of their subject. In some cases this effort was applied in a haphazard manner, but in others people realized the immense potential of using computers to model and analyse their data. This small band of biologists along with a handful of interested computer scientists, mathematicians, crystallographers, and physical scientists (in no particular order of priority or importance), formed the fledgling bioinformatics community. It has been a unique feature of the field that the most useful and exciting work has been carried out as collaborations between researchers from these different disciplines.

By 1985, there was the first journal devoted (largely or partly) to the subject: Computer Applications in the Biosciences. Bioinformatics articles tended to dominate it and the name was changed to reflect this, a few years ago when it was re-christened, simply, as Bioinformatics. By then, the EMBL sequence data library in Heidelberg had been running for four years, followed closely by the US based, GenBank. The first releases of the DNA sequence databases were sent out as printed booklets as well as on computer tapes. It was routine to simply dump the tape contents to a printer anyway as computer disk space in those days was expensive. This practice became pointless and impossible by 1985, due to the speed with which DNA sequence data were accumulating.

During the 1990s, the entire field of bioinformatics was transformed, almost beyond recognition by a series of developments. Firstly, the internet became the standard computer network world wide. Now, all new analyses, services, data sets, etc. could be made available to researchers across the world by a simple annoucement to a bulletin board/newsgroup and the setting up of a few pages on the World Wide Web (WWW). Secondly, advances in sequencing technology have made it almost routine to think in terms of sequencing the entire genome of organisms of interest. The generation of genome data is a completely computer-dependent task; the interpretation is impossible without computers and to access the data you need to use a computer. Bioinformatics has come of age.

Sequence analysis and searching

Since the first efforts of Gilbert and Sanger, the DNA sequence databases have been doubling in size (numbers of nucleotides or sequences) every 18 months or so. This trend continues unabated. This forced the development of systems of software and mathematical techniques for managing and searching these collections. Earlier, the main labs generating the sequence data in the first place had been forced to develop software to help assemble and manage their own data. The famous Staden package came from work by Roger Staden in the LMB in Cambridge (UK) to assemble and analyse data from the early DNA sequencing work in the laboratory of Fred Sanger.

The sheer volume of data made it hard to find sequences of interest in each release of the sequence databases. The data were distributed as collections of flat files, each of which contained some textual information (the annotation) such as organism name and keywords, as well as the DNA sequence. The main way of searching for sequences of interest was to use a string-matching program or to browse a printout of some annotation by hand. This forced the development of relational database management systems in the main database centres but the databases continued to be delivered as flat files. One important early system, that is still in use, for browsing and searching the databases, was ACNUC, from Manolo Gouy and colleagues in Lyon, France. This was developed in the mid-eighties and allowed fully relational searching and browsing of the data base annotation. SRS is a more recent development and is described fully in Chapter 10 of this volume.

A second problem with data base size was the time and computational effort required to search the sequences themselves for similarity with a search sequence. The mathematical background to this problem had been worked on over the 1970s by a small group of mathematicians and the gold standard method was the well-known Smith and Waterman algorithm, developed by Michael Waterman (a mathematician) and Temple Smith (a physicist). The snag was that computer time was scarce and expensive and it could take hours on a large mainframe to carry out a typical search. In 1985, the situation changed dramatically with the advent of the FASTA program. FASTA was developed by David Lipman and Bill Pearson (both biologists in the US). It was based on an earlier method by John Wilbur and Lipman which was in turn based on an earlier paper by two Frenchmen (Dumas and Ninio) who showed how to use standard techniques from computer science (linked lists and hashing) to quickly compare chunks of sequences. FASTA caused a revolution. It was cheap (basically free), fast (typical searches took just a few minutes), and ran on the newly available PCs (personal computers). Now, biologists everywhere could do their own searches and do them as often as they liked. It became standard practice, in laboratories all over the word, to discover the function of newly sequenced genes by carrying out FASTA searches of databases of characterized proteins. Fortunately, by this time the databases were just big enough to give some chance of finding a similar sequence in a search with a randomly chose gene. Sadly, the chances were small initially, but by the early nineties they had risen to 1 in 3 and now are well over 50%.

By 1990, even FASTA was too slow for some types of search to be carried out routinely, but this was alleviated by the development of faster and faster workstations. A parallel development was the use of specialist hardware such as super-computers or massively parallel computers. These allowed Smith and Waterman searches to be carried out in seconds and one very successful service was provided by John Collins and Andrew Coulson in Edinburgh, UK. The snag with these developments was the sheet cost of these specialist computers and the great skill required to write the computer code so networks were important. If you could not afford a big fast box of specialized chips, you might know someone who would allow you to use theirs and you could log on to it using a computer network.

In 1990, a new program called BLAST appeared. It was written by a collection of biologists, mathematicians and computer scientists, mainly at the new NCBI, in Washington DC, USA. It filled a similar niche to the FASTA program but was an order of magnitude faster for many types of search. It also featured the use of a probability calculation in order to help rank the importance of the sequences that were hit in the search (see Chapter 8 for some details). Probability calculations are now very important in many areas of bioinformatics (such as hidden Markov models; see chapter 4).

Protein structure analysis and prediction

Protein structure plays a central role in our understanding and use of sequence data. A knowledge of the protein structure behind the sequences often makes clear what mutational constraints are imposed on each position in the sequence and can therefore aid in the multiple alignment of sequences (Chapters 1, 3, and 6) and the interpretaion of sequence patterns (Chapter 7). While computational methods have been developed for comparing sequences with sequences (which, as we have already seen, are critical in databank searching), methods have also been developed for comparing sequences with structures (something called 'threading') and structures with structures (Covered in Chapters 1 and 2, respectively). All these methods support each other and roughly following the progression: (1) DB-search → (2) multiple alignment → (3) threading → (4) modelling. However, this is often far from a linear progression: the alignment can reveal new constraints that can be imposed on the databank search, while at the same time also helping the threading application. Similarly, the threading can cast new (structural) light on the alignment and all are carried out under (and also affect) the prediction of secondary structure.

Before the advent of multiple genome data, this favoured route often came to a halt before it started: when no similar sequence could be found even to make an alignment. However, with the genomes of phylogentically widespread organisms either completed or promised soon (bacteria, yeast, plasmodium, worm, fly, fish, man) there is now a good chance of finding proteins from each that can compile a useful multiple-sequencing alignment. At the threading stage (2) in the above progression, the current problem and worry is that there may not be a protein structure on which the alignment can be fitted. Failiure at this stage generally compromises any success in the final modelling stage (unless sufficient structural constraints are available from other experimental sources). This problem will be eased by structural genomics programmes (often associated with a genome program) for the large-scale determination of protein structures. As with the genome, these data will greatly increase the chance of finding at least one structure onto which the protein can be modelled.

The future of a mature field

With several complete genomes and a reasonably complete set of protein structures, the problems facing Bioinformatics shifts from its past challenge of finding weak similarities among sparse data, to one of finding closer similarities in a wealth of data. However, concentrating on protein sequence data (as distinct from the raw genomic DNA) eases the data processing problem considerably and the increased computation demands can be met by the equally rapid increase in the power of computers. In this new situation, perhaps all that will be needed is a good multiple sequence alignment program (such as CLUSTAL or MULTAL) with which to reveal all necessary functional and structural information on any particular gene.

The most fundamental impact of the 'New Data' is the realization that the biological world is finite and, at least in the world of sequences, that we have the end in sight. We have already, in the many bacterial genomes and in yeast, seen the minimal complement of proteins required to maintain independent life— and at only several thousand proteins, it does not seem unworkably large. This will expand by an order-or-magnitude in the higher organisms but it is already clear that much of this expansion can be accounted for by the proliferation of sequences within tissue or functionally specific families (such as the G-protein coupled receptors). Removing this 'redundancy' might still result in a set of proteins that, if not by eye, can be easily analysed by computer.

The end-of-the-line in protein structures may take a little longer to arrive, but, by implication from the sequences, it too is finite—and indeed, may be much more finite than the sequence world. This can be inferred from current data by the number of protein families that have the same overall structure (or fold), but otherwise exhibit no signs of functional or sequence similarity. Besides comparing and classifying the different structures, an interesting aspect is to develop models of protein structure evolution, perhaps allowing very distant relationships between these different folds to be inferred. It might be hoped that this will shed light on the most ancient origins of protein structure and on the distant relationships between biological systems.

The ultimate aim of Bioinformatics must surely be the complete understanding of an organism—given its genome. This will require the characterization and modelling of extremely complex systems: not only within the cell but also including the fantastic network of cell–cell interactions that go to make-up an organism (and how the whole system boot-straps itself). However, as Sergeant Pluck has told us: what is an organism but only millions of little bits of itself whirling around and doing intricate convolutions. If a genome can tell us all these bits (and sure it will be no time till we have the genome for a sheep) then all we have to do is figure out how it all whirls around. For this, without a doubt, the Sergeant would have recommended the careful application of algebra—and, had he known about them, I'm sure he would have used a computer.

D.H. and W.T., 2000

Contents

Preface *page* *v*

List of protocols *xvii*

Abbreviations *xix*

1 Threading methods for protein structure prediction *1*

David Jones and Caroline Hadley

1 Introduction *1*

2 Threading methods *1*

　1-D–3-D profiles: Bowie *et al.* (1991) *5*

　Threading: Jones *et al.* (1992) *5*

　Protein fold recognition using secondary structure predictions: Rost (1997) *7*

　Combining sequence similarity and threading: Jones (1999) *7*

3 Assessing the reliability of threading methods *8*

　Alignment accuracy *9*

　Post-processing threading results *10*

　Why does threading work? *10*

4 Limitations: strong and weak fold-recognition *11*

　The domain problem in threading *11*

5 The future *12*

　References *12*

2 Comparison of protein three-dimensional structures *15*

Mark S. Johnson and Jukka V. Lehtonen

1 Introduction *15*

2 The comparison of protein structures *16*

　General considerations *16*

　What atoms/features of protein structure to compare? *17*

　Standard methods for finding the translation vector and rotation matrix *20*

　Standard methods to determine equivalent matched atoms between structures *25*

　Quality and extent of structural matches *29*

3 The comparison of identical proteins *31*

　Why compare identical proteins? *31*

　Comparisons *31*

4 The comparison of homologous structures: example methods *32*
  Background *32*
  Methods that require the assignment of seed residues *34*
  Automatic comparison of 3-D structures *35*
  Multiple structural comparisons *41*

5 The comparison of unrelated structures *42*
  Background *42*

6 Large-scale comparisons of protein structures *46*
  References *48*

3 Multiple alignments for structural, functional, or phylogenetic analyses of homologous sequences *51*

L. Duret and S. Abdeddaim

1 Introduction *51*

2 Basic concepts for multiple sequence alignment *53*
  Homology: definition and demonstration *53*
  Global or local alignments *54*
  Substitution matrices, weighting of gaps *54*

3 Searching for homologous sequences *56*

4 Multiple alignment methods *57*
  Optimal methods for global multiple alignments *59*
  Progressive global alignment *61*
  Block-based global alignment *63*
  Motif-based local multiple alignments *65*
  Comparison of different methods *65*
  Particular case: aligning protein-coding DNA sequences *68*

5 Visualizing and editing multiple alignments *69*
  Manual expertise to check or refine alignments *71*
  Annotating alignments, extracting sub-alignments *71*
  Comparison of alignment editors *72*
  Alignment shading software, pretty printing, logos, etc. *72*

6 Databases of multiple alignments *72*

7 Summary *73*
  References *74*

4 Hidden Markov models for database similarity searches *77*

Ewan Birney

1 Introduction *77*

2 Overview *78*

3 Using profile and profile-HMM databases *79*
  Pfam *80*
  Prosite profiles *80*
  SMART *81*
  Other resources and future directions *81*
  Limitations of profile-HMM databases *81*

4 Using PSI-BLAST *81*

5 Using HMMER2   *82*
  Overview of using HMMER   *83*
  Making the first alignment   *83*
  Making a profile-HMM from an alignment   *84*
  Finding homologues and extending the alignment   *84*

6 False positives   *85*

7 Validating a profile-HMM match   *85*

8 Practical issues of the theories behind profile-HMMs   *86*
  Overview of profile-HMMs   *86*
  Statistics for profile-HMM   *87*
  Profile-HMM construction   *89*
  Priors and evolutionary information   *89*
  Technical issues   *90*

  References   *91*

5 Protein family-based methods for homology detection and analysis   *93*

Steven Henikoff and Jorja Henikoff

1 Introduction   *93*
  Expanding protein families   *93*
  Terms used to describe relationships among proteins   *93*
  Alternative approaches to inferring function from sequence alignment   *94*

2 Displaying protein relationships   *95*
  From pairwise to multiple-sequence alignments   *95*
  Patterns   *96*
  Logos   *97*
  Trees   *97*

3 Block based methods for multiple sequence alignment   *98*
  Pairwise alignment-initiated methods   *98*
  Pattern-initiated methods   *99*
  Iterative methods   *99*
  Implementations   *100*

4 Position-specific scoring matrices (PSSMs)   *101*
  Sequence weights   *102*
  PSSM column scores   *102*

5 Searching family databases with sequence queries   *103*
  Curated family databases: Prosite, Prints, and Pfam   *105*
  Clustering databases: ProDom, DOMO, Protomap, and Prof_pat   *105*
  Derived family databases: Blocks and Proclass   *106*
  Other tools for searching family databases   *107*

6 Searching with family-based queries   *108*
  Searching with embedded queries   *108*
  Searching with PSSMs   *108*
  Iterated PSSM searching   *109*
  Multiple alignment-based searching of protein family databases   *110*

  References   *110*

6 Predicting secondary structure from protein sequences *113*

Jaap Heringa

1 Introduction *113*
    What is secondary structure? *113*
    Where could knowledge about secondary structure help? *114*
    What signals are there to be recognized? *114*

2 Assessing prediction accuracy *118*

3 Prediction methods for globular proteins *120*
    The early methods *120*
    Accuracy of early methods *122*
    Other computational approaches *122*
    Prediction from multiply-aligned sequences *123*
    A consensus approach: JPRED *129*
    Multiple-alignment quality and secondary-structure prediction *131*
    Iterated multiple-alignment and secondary structure prediction *132*

4 Prediction of transmembrane segments *133*
    Prediction of α-helical TM segments *134*
    Orientation of transmembrane helices *136*
    Prediction of β-strand transmembrane regions *136*

5 Coiled-coil structures *137*

6 Threading *138*

7 Recommendations and conclusions *138*
    References *139*

7 Methods for discovering conserved patterns in protein sequences and structures *143*

Inge Jonassen

1 Introduction *143*

2 Pattern descriptions *144*
    Exact or approximate matching *144*
    PROSITE patterns *145*
    Alignments, profiles, and hidden Markov models *146*
    Pattern significance *148*
    Pattern databases *150*
    Using existing pattern collections *153*

3 Finding new patterns *154*
    A general approach *154*
    Discovery algorithms *155*

4 The Pratt programs *156*
    Using Pratt *157*
    Pratt: Internal search methods *159*
    Scoring patterns *161*

5 Structure motifs *162*
    The SPratt program *162*

6 Examples *164*

7 Conclusions *164*
    References *165*

8 Comparison of protein sequences and practical database searching *167*

Golan Yona and Steven E. Brenner

1 Introduction *167*

2 Alignment of sequences *168*
  Rigorous alignment algorithms *169*
  Heuristic algorithms for sequence comparison *171*

3 Probability and statistics of sequence alignments *173*
  Statistics of global alignment *174*
  Statistics of local alignment without gaps *175*
  Statistics of local alignment with gaps *177*

4 Practical database searching *178*
  Types of comparison *178*
  Databases *179*
  Algorithms *181*
  Filtering *181*
  Scoring matrices and gap penalties *182*
  Command line parameters *185*

5 Interpretation of results *187*

6 Conclusion *188*

  References *188*

9 Networking for the biologist *191*

R. A. Harper

1 Introduction *191*

2 The changing face of networking *192*
  Networking in Europe *194*
  The way we were . . . e-mail servers for sequence retrieval *195*
  Similarity searches via e-mail *199*
  Speed solutions for similarity searches *201*

3 Sequence retrieval via the WWW *203*
  Entrez from the NCBI *205*
  SRS from the EBI *205*

4 Submitting sequences *208*
  Bankit at NCBI *209*
  Sequin from NCBI *209*
  Webin from EBI *210*
  Sakura from DDJB *212*

5 Conclusions *212*

  References *213*

10 SRS—Access to molecular biological databanks and integrated data analysis tools *215*

D. P. Kreil and T. Etzold

1 Introduction *215*
  SRS fills a critical need *215*
  History, philosophy, and future of SRS *216*

2 A user's primer   *217*
    A simple query   *219*
    Exploiting links between databases   *220*
    Using Views to explore query results   *221*
    Launching analysis tools   *223*
    Overview   *225*

3 Advanced tools and concepts   *225*
    Refining queries   *225*
    Creating custom Views   *230*
    SRS world wide: using DATABANKS   *232*
    Interfacing with SRS over the network   *233*

4 SRS server side   *236*
    User's point of view   *236*
    Administrator's point of view   *238*

5 Where to turn to for help   *240*

Acknowledgements   *241*

References   *241*

List of suppliers   *243*

Index   *247*

Protocol list

The comparison of protein structures

Features used for the comparison of protein 3-D structures   *19*
Rigid-body structural comparisons: translations and rotations   *22*
The alignment: determination of equivalent pairs   *27*
Root mean squared deviations (RMSD)   *30*

The comparison of identical proteins

Similarities among different structures of identical proteins   *32*

The comparison of homologous structures: example methods

Finding initial seed residues   *34*
Semi-automatic methods   *34*
Structural comparisons seeded from sequence alignments   *35*
GA_FIT (ref. 26, 27)   *36*
Local similarity search by VERTAA   *37*
Structure comparison by DALI (11)   *39*
Structure comparison by SSAP (10)   *40*
Structure comparison by DEJAVU (30)   *40*
Multiple structural alignments from pairwise comparisons   *41*

The comparison of unrelated structures

Structure comparison by SARF2 (41)   *45*
GENFIT (22)   *45*

Block-based methods for multiple-sequence alignment

Finding motifs from unaligned sequences and searching sequence databanks   *100*

Assessing prediction accuracy

Jackknife testing   *119*

Recommendations and conclusions

Predicting secondary structure   *139*

The changing face of networking

Using *netserv@ebi.ac.uk* for the retrieval of sequences and software   *198*
Using *query@ncbi.nlm.nih.gov* for sequence retrieval   *198*
Blast similarity search e-mail server at NCBI   *199*
FASTA similarity search e-mail server at EBI   *201*

A user's primer

Performing a simple SRS query   *219*

Applying a link query to selected entries   *220*

Displaying selected entries with one of the pre-defined views   *221*

Launching an external application program for selected entries   *223*

Advanced tools and concepts

Browsing the index for a database field   *227*

Search SRS world wide   *232*

Abbreviations

AACC	amino acid class covering
API	application programming interface
CASP	critical assessment in structure prediction
3-D	three-dimensional
EBI	European Bioinformatics Institute
EM	expectation-maximization
e-value	expectation value
EVD	extreme value distribution
FDF	fast data finder
FN	false negative
HMM	hidden Markov model
ILP	inductive logic programming
LAMA	local alignment of multiple alignments
MAST	multiple alignment searching tool
MDL	minimum description length
MP	membrane protein
NCBI	National Centre for Biotechnology Information
NCGR	National Centre for Genomic Research
NNSSP	nearest neighbour secondary structure prediction
PD	pattern driven
PDB	protein data bank
PHD	profile secondary structure predictions from Heidelberg
PPV	positive predictive value
PSSM	position-specific scoring matrices
QOS	quality of service
RMSD	root mean square deviation
RTT	round trip time
SAP	structure alignment program
SD	sequence driven
SP	sum of pairs
SRS	sequence retrieval system

SSE	secondary structure element
TM	transmembrane
TP	true positive
WWW	World Wide Web

Chapter 1

Threading methods for protein structure prediction

David Jones and Caroline Hadley

Department of Biological Sciences, University of Warwick, Coventry CV4 7AL, UK.

1 Introduction

As the attempts to sequence entire genomes increases the number of protein sequences by a factor of two each year, the gap between sequence and structural information stored in public databases is growing rapidly. In stark contrast to sequencing techniques, experimental methods for structure determination are time-consuming, and limited in their application, and therefore will not be able to keep pace with the flood of newly characterized gene products. The development of practical methods for predicting protein structure from sequence is therefore of considerable importance in the field of biology.

Several different approaches have been used to predict protein structure from sequence, with varying degrees of success. *Ab initio* methods encompass any means of calculating co-ordinates for a protein sequence from first principles—that is, without reference to existing protein structures. Little success has been seen in this area, with more theory produced than actual useful methodology. Comparative (or homology) modelling, attempts to predict protein structure on the strength of a protein's sequence similarity to another protein of known structure (following the theory that similar sequence implies similar structure). Some success has been achieved, but several limitations to this method, not least of which are its dependence on alignment quality and the existence of a good sequence homologue, indicate it is not applicable to a large fraction of protein sequences. The third main category of protein structure prediction, falling somewhere between comparative modelling and *ab initio* prediction, is fold recognition, or threading.

2 Threading methods

The term 'threading' was first coined in 1992 by Jones *et al.* (1), but the field has grown considerably since then with many different methods being proposed: for example, Godzik and Skolnick (2); Ouzounis *et al.* (3); Abagyan *et al.* (4); Overington *et al.* (5); Matsuo *et al.* (6); Madej *et al.* (7); Lathrop and Smith (8);

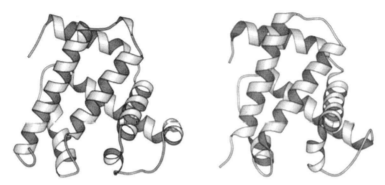

Figure 1 An example of a pair of protein structures in the same family. (a) Human myoglobin [2mm1], (b) pig haemoglobin, alpha chain [2pghA]. At the family level, proteins have higher sequences identity (in this case, 32%) and have highly similar structures. Figures created using Molscript (19).

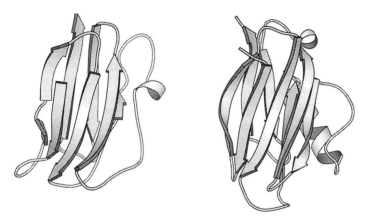

Figure 2 A pair of structures within the same superfamily. (a) *A. denitrificans* azurin [1azcA], (b) poplar plastocyanin [1plc]. Members of the same superfamily may have insignificant sequence identity (16% in this case), but still share most features of the protein fold, reflecting a common evolutionary origin.

Taylor (9) amongst others. The idea behind threading came about from the observation that a large percentage of proteins adopt one of a limited number of folds (*Figures 1–3*). In fact, just 10 different folds (the 'superfolds') account for 50% of the known structural similarities between protein superfamilies (18). Thus, rather than trying to find the correct structure for a protein from the huge number of all possible conformations available to a polypeptide chain, the correct (or close to correct) structure is likely to have already been observed and already stored in a structural database. Of course, in cases where the target protein shares significant sequence similarity to a protein of known 3-D structure, the 'fold recognition' problem is trivial—simple sequence comparison will identify the correct fold. The hope was, however, that threading might be able to detect structural similarities that are not accompanied by any detectable sequence similarity, and this has subsequently been proven to be the case.

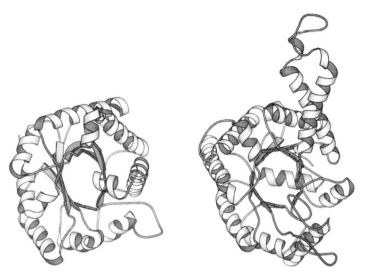

Figure 3 A pair of analogous folds. (a) Chicken triosephosphate isomerase [1timA], (b) *E. coli* fructose bisphosphate aldolase [1dosA]. Members of the same fold family have the same major secondary structure elements with the same arrangement and connectivity. Very low sequence identity and large variations in the details of the structures reflects the lack of common ancestry between analogous folds. Although the aldolase structure contains additional helices, the TIM barrel fold is obviously present in both proteins. The TIM barrel is one of 10 'superfolds' identified by Orengo *et al.* (18).

Figure 4 shows an outline of a generic fold recognition method. Firstly, a library of unique or representative protein structures needs to be derived from the database of all known protein structures. Different groups use different selection criteria for their fold libraries: in some cases, complete protein chains are used in the library, but in other cases, structural domains or even conserved proteins cores are used. Each fold from this library is then considered in turn and the target sequence optimally fitted (or aligned) to each library fold (allowing for relative insertions and deletions in loop regions). Many different algorithms have been proposed for finding this optimal sequence–structure alignment, with most groups using some form of dynamic programming algorithm (including the examples described below), but other algorithms such as Gibbs sampling (7) or branch-and-bound searching (8) have also been used with some success. Finally, some kind of objective function is needed to determine the goodness of fit between the sequence and the template structure. It is this objective function which is optimized during the sequence–structure alignment. Again opinions differ as to the form of this objective function. Most groups use some kind of 'pseudo energy' function based on a statistical analysis of observed protein structures, but other more abstract scoring functions have also been proposed (see ref. 20 for a recent review). The final result of a fold recognition method is a ranking of the fold library in descending order of 'goodness of fit', with the best fitting fold (typically the lowest energy fold) being taken as the most probable match.

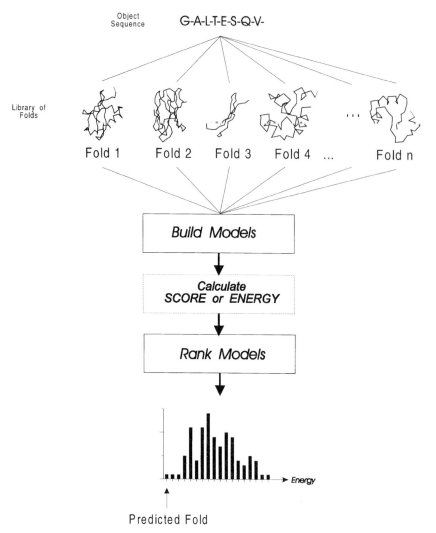

Object Sequence G-A-L-T-E-S-Q-V-

Library of Folds

Fold 1   Fold 2   Fold 3   Fold 4   ...   Fold n

Build Models

Calculate SCORE or ENERGY

Rank Models

Energy

Predicted Fold

Figure 4 This is an outline of the fold recognition approach to protein structure prediction, and identifies three clear aspects of the problem that need consideration: a fold library, a method for modelling the object sequence on each fold, and a means for assessing the goodness-of-fit between the sequence and the structure.

No matter what algorithm or scoring function is used, fold recognition is not without its limitations, and some progress must be made before it can be considered a routine protein structure prediction tool. Several different aspects of this method are particularly open to improvement, namely the question of potential functions (i.e. the calculations used to determine the energy of a particular sequence once fitted onto a template fold), improvements in alignments (i.e. correctly aligning the sequence onto the template fold, to produce the best fit), and the need for progress in post-processing the results (i.e. from the energy calculations, etc., choosing the best 'fit'). Significant progress may also arise from improvements in the threading library used (i.e. the templates upon which the sequences will be threaded).

To get some idea of the variety of methods which have been developed, four distinct approaches to the fold-recognition problem will be described. Virtually all fold-recognition methods are similar to at least one of these methods, and some newer methods incorporate concepts from more than one.

2.1 1-D–3-D profiles: Bowie *et al.* (1991)

The first true fold recognition method was by Bowie, Lüthy, and Eisenberg (10), where they attempted to match sequences to folds by describing the fold in terms of the *environment* of each residue in the structure. The environment was described in terms of local secondary structure (3 states: α, β, and coil), solvent accessibility (3 states: buried, partially buried, and exposed), and the degree of burial by polar rather than apolar atoms. The basic idea of the method is the assumption that the environment of a particular residue thus defined is expected to be more conserved than the actual residue itself, and so the method is able to detect more distant sequence–structure relationships than purely sequence-based methods. The authors describe this method as a 1-D–3-D profile method, in that a 3-D structure is translated into a 1-D string, which can then be aligned using traditional dynamic programming algorithms. Bowie *et al.* have applied the 1-D–3-D profile method to the inverse folding problem and have shown that the method can indeed detect remote matches, but in the cases shown the hits still retained some weak sequence similarity with the search protein. Environment-based methods appear to be incapable of detecting structural similarities between extremely divergent proteins, and between proteins sharing a common fold through convergent evolution—environment only appears to be conserved up to a point. Consider a buried polar residue in one structure that is found to be located in a polar environment. Buried polar residues tend to be functionally important residues, and so it is not surprising then that a protein with a similar structure but with an entirely different function would choose to place a hydrophobic residue at this position in an apolar environment. A further problem with environment-based methods is that they are sensitive to the multimeric state of a protein. Residues buried in a subunit interface of a multimeric protein will not be buried at an equivalent position in a monomeric protein of similar fold.

2.2 Threading: Jones *et al.* (1992)

The method which introduced the term 'threading' (1) went further than the method of Bowie, Lüthy, and Eisenberg in that instead of using averaged residue environments, a given protein fold was modelled in terms of a 'network' of pairwise interatomic energy terms, with the structural role of any given residue described in terms of its interactions. Classifying such a set of interactions into one environmental class such as 'buried alpha helical' will inevitably result in the loss of useful information, reducing the *specificity* of sequence–structure matches evaluated in this way. Thus, in true threading methods, a sequence is matched to a structure by considering detailed pairwise interactions, rather than averaging them into a crude environmental class. However, incorporation of such non-local interactions means that simple dynamic programming string-matching methods cannot be used. There is therefore a trade-off to be made between the complexity of the sequence–structure scoring scheme and the algorithmic complexity of the problem.

Jones *et al.* (1) proposed a novel dynamic programming algorithm (now commonly known as 'double' dynamic programming) to the problem of aligning a given sequence with the backbone co-ordinates of a template protein structure, taking into account the detailed pairwise interactions. The problem of matching pairwise interactions is somewhat similar to the problem of structural comparison methods. The *potential environment* of a residue i can be defined as being the sum of all pairwise potential terms involving i and all other residues $j{\neq}i$. This is an analogous definition to that of a residue's *structural environment*, as described by Taylor and Orengo (11). In the simplest case, structural environment of a residue i may be defined as the set of all inter-Cα distances between residue i and all other residues $j{\neq}i$. Taylor and Orengo propose a novel dynamic programming algorithm for the comparison of such residue structural environments, and this method proved to be effective for the comparison of residue potential environments. A detailed description of the algorithm has recently been published (12).

For a sequence–structure compatibility function, Jones *et al.* chose to use a set of statistically derived pairwise potentials similar to those described by Sippl (13). Using the formulation of Sippl, short (sequence separation, $k \leq 10$), medium ($11 \leq k \leq 30$), and long ($k > 30$) range potentials were constructed between the following atom pairs: C$\beta \rightarrow$ Cβ, C$\beta \rightarrow$ N, C$\beta \rightarrow$ O, N $\rightarrow$ Cβ, N $\rightarrow$ O, O $\rightarrow$ Cβ, and O $\rightarrow$ N. For a given pair of atoms, a given residue sequence separation and a given interaction distance, these potentials provide a measure of energy, which relates to the probability of observing the proposed interaction in native protein structures. In addition to these pairwise terms, a 'solvation potential' was also incorporated. This potential simply measures the frequency with which each amino acid species is found with a certain degree of solvation, approximated by the residue solvent accessible surface area.

By dividing the empirical pair potentials into sequence separation ranges, specific structural significance may be tentatively conferred on each range. For instance, the short range terms predominate in the matching of secondary structural elements. By threading a sequence segment onto the template of an alpha helical conformation and evaluating the short range potential terms, the possibility of the sequence folding into an alpha helix may be evaluated. In a similar way, medium range terms mediate the matching of super-secondary structural motifs, and the long range terms, the tertiary packing.

Recent features added to the method allow sequence information and predicted secondary structure information to be considered in the fold-recognition process. Sequence information is weighted into the fold recognition potentials using a transformation of a mutation data matrix (12). By carefully selecting the weighting of the sequence components in the scoring function it is possible to balance the influence of sequence matching with the influence of the pairwise and solvation energy terms. In contrast to this, secondary structure information is not incorporated into the sequence–structure scoring function. In this case, secondary structure information is used to mask regions of the alignment path matrix so that the threading alignments do not align (for example) predicted β strands with observed α helices. A confidence threshold is applied to the secondary structure prediction data so that only the most confidently predicted regions of the prediction are used to mask the alignment matrix.

2.3 Protein fold recognition using secondary structure predictions: Rost (1997)

Although most fold recognition methods employ potentials of one kind or another, it is quite easy to design a useful fold recognition approach that at first sight does not employ potentials of any kind. Although not the first example of this approach, as a good recent example the PHD secondary structure prediction service (14) has recently been extended to offer a fold recognition option. In this case the system predicts the secondary structure and accessibility of each residue in the protein of interest, encodes this information in the form of a string (similar to the scheme employed by Bowie *et al.*) (10) and then matches this string against a library of strings computed from known structures. A number of other similar methods are also in development in other labs, though all based on the initial prediction of secondary structure by PHD. Clearly no explicit potentials are being employed in these methods, but potentials are implicitly coded into the neural network weights used to predict secondary structure in the first place.

2.4 Combining sequence similarity and threading: Jones (1999)

Jones (15) has recently proposed a hybrid fold recognition method which is designed to be both fast and reliable, and is particularly aimed at automated genome annotation. The method uses a sequence profile-based alignment algorithm to generate alignments which are then evaluated by threading techniques. As a last step, each threaded model is evaluated by a neural network in order to produce a single measure of confidence in the proposed prediction. The speed of the method, along with its sensitivity and very low false-positive rate makes it ideal for automatically predicting the structure of all the proteins in a translated bacterial genome. The method has been applied to the genome of *Mycoplasma genitalium*, and analysis of the results shows that as many as 46% (now 51%) of the proteins derived from the predicted protein coding regions have a significant relationship to a protein of known structure. The fact that alignments are generated by a sequence alignment step means that the method

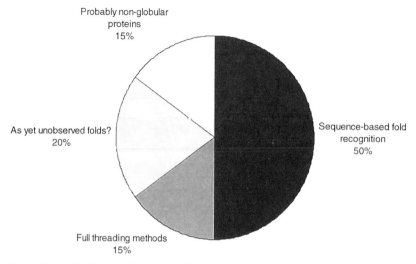

Probably non-globular
proteins
15%

As yet unobserved folds?
20%

Sequence-based fold
recognition
50%

Full threading methods
15%

Figure 5 Hypothetical applicability of different categories of fold-recognition methods to the Open Reading Frames of small bacterial genomes. At present sequence-based fold recognition (e.g. GenTHREADER) is successful for around 50% of the ORFs. Structures for a further 15% of ORFs can probably be assigned by full threading methods such as THREADER, and the remaining 35% cannot currently be recognized either because the fold has not yet been observed, or because the ORF encodes a non-globular protein (e.g. a transmembrane protein).

is only expected to work for family or superfamily level similarities between the target and template proteins. This is both a positive and negative feature of the method. The negative aspect is that, of course, many purely structural similarities will not be detected by the method. The positive aspect is that superfamily relationships produce the most reliable results, and also allow some aspects of the function of the target protein to be inferred from the matched template structure. This latter point is particularly useful when annotating unknown genome sequences. *Figure 5* shows the current applicability of different types of fold recognition method to a genome such as that of *M. genitalium*.

Unlike full threading methods, which require a great deal of computer power to run, this type of method can be made readily available to the public via a simple Web server. The GenTHREADER method is available from the following URL:

http://globin.bio.warwick.ac.uk/psipred

3 Assessing the reliability of threading methods

Although the published results for the fold recognition methods can look impressive, showing that threading is indeed capable of recognizing folds in the absence of significant sequence similarity, it can be argued that in all cases the correct answers were already known and so it is not clear how well they would perform in real situations where the answers are not known at the time the predictions are made. It was not until these methods were tested in a set of blind trials—the Critical Assessment in Structure Prediction experiments (CASP)—that it became clear how powerful these methods could be when used without prior knowledge of the correct answer. The CASP experiment has now been run three times (CASP1 in 1994, CASP2 in 1996, CASP3 in 1998): and in the last meeting results from over 30 methods were evaluated by the independent assessors. Up to date information on all of the CASP experiments can be obtained from the following Web address:

http://predictioncenter.llnl.gov

3.1 Alignment accuracy

Most published methods are evaluated solely on the basis of fold assignment, i.e. the method is evaluated on its ability to correctly pick the correct fold. However, in practice, fold assignment is not sufficient in its own right. Given a correct fold assignment the next step is of course to generate an accurate sequence structure alignment and to use this alignment to generate an accurate 3-D model for the target protein. In cases where a fold has been assigned, the alignment can be passed to an automatic comparative modelling program (e.g. MODELLER3) so that loops and side chains can be built in.

The accuracy of alignments that can be produced by fold recognition methods can be measured in terms of the Root Mean Square Deviation (RMSD) between the implied prediction model and the observed experimental structure. Analysis of the results of the CASP experiments has shown that alignment accuracy correlates strongly with the degree of evolutionary and structural divergence between the available template structures and the target protein. The degree of model accuracy that can be expected can be broken down into three categories of structure relationship:

(a) **Family** (e.g. *Figure 1*). Evident sequence similarity. Threading models will be almost entirely accurate, with an RMSD of between 1.0 and 3.0 Angstroms, depending on the degree of sequence similarity.

(b) **Superfamily** (e.g. *Figure 2*). No significant sequence similarity, but evident common ancestry between the template and target structure. Models for this class of similarity will be partially correct (mostly in active site regions) and will have an RMSD of between 3.0 and 6.0 Angstroms typically (though sometimes more depending on the accuracy of the alignment produced).

(c) **Analogy** (e.g. *Figure 3*). No apparent common ancestry between the template and target structure. Low quality models are expected for this category of similarity. RMSD is not a good way to evaluate models of this quality as very large shifts in the alignment produce virtually random RMSD values. At best, alignments in this class are 'topologically correct', in that the correct elements of secondary structure are equivalenced, but frequently shifts in the alignment are so large as to render the models entirely incorrect.

3.2 Post-processing threading results

Perhaps one of the most significant observations that came from the CASP2 prediction experiment was that a great deal of success in fold recognition can be achieved purely from a deep background knowledge of protein structure and function relationships. Alexey Murzin (one of the authors of the SCOP protein structure classification scheme) identified a number of key evolutionary clues which led him to correctly assign membership of some of the target proteins to known superfamilies (16). Also in two cases he was able to confidently assign a 'null prediction' to targets with unique folds purely by considering their predicted secondary structure. These feats are quite remarkable, but not easily reproduced by non-experts in protein structure and function. Despite this, it is very clear that an important future development of practical fold-recognition is to take both structure and function into account when ranking the sequence–structure matches.

Even without new developments in fold recognition algorithms, information on function and other sources of information can be applied to the results of a threading method as a 'post-processing' step. Rather than simply taking the top scoring fold to be the assumed correct answer, a fold from, say, the top 10 matches can be selected by human intervention. Such intervention might involve visual inspection of the proposed alignment, inspection of the proposed 3-D structure on a graphics workstation, comparison of proposed secondary structure with that obtained from secondary structure prediction or even consideration of common function between the target and template proteins.

3.3 Why does threading work?

Although many different formulations of energy function have been used for fold recognition, it has been shown that the principal factor in the most successful of these empirical potentials essentially encodes the general 'hydrophobic effect', rather than specific interactions between specific side chains. (e.g. the interaction potential between like charges is the same as that derived for unlike charges, reflecting not the specific interaction between side chains, but their overall preference to lie on the surface). Despite this observation that specific pair interactions are not vital to successful fold recognition, threading methods based on pairwise interactions do seem to work better than profile methods (as evidenced for example in the predictions made during the CASP experiments). This might at first sight seem contradictory, particularly as it is apparent that specific pairwise interactions are not conserved between analogous fold families (17). Nevertheless, threading methods do seem to be picking up signals which are not detected by simple 1-D profile methods. Why might this be the case?

One reasonable explanation may be that profile-based fold-recognition methods make the assumption that the pattern of accessibility between two divergent protein structures is perfectly conserved, and it is this assumption that results in their relatively poor performance. Threading methods, on the other hand, are able to model the environment of a residue by summing the hydrophobic pair interactions surrounding a particular residue. These pair interaction environments of course change as the threading alignment changes, and it is this sensitivity of residue environments to changes in the sequence–structure alignment that results in the increased predictive power of threading methods. Although this explains why threading works even when specific contacts are not being conserved, it also explains why sequence–structure alignments are generally of poor quality when compared with known structure–structure alignments.

4 Limitations: strong and weak fold-recognition

What are the limitations of current fold-recognition methods? Let's consider two forms of the fold-recognition problem. In the first form of the problem we seek a set of potentials (and a method for performing the sequence–structure alignment) which will reliably recognize the closest matching fold for a given sequence from the thousands of alternatives—as many as 7000 naturally occurring folds have been estimated (18). This form of the problem is referred to as the 'strong' fold-recognition problem. It is possible that the strong fold-recognition problem is actually insoluble because, quite simply, the real protein free energy function is itself almost certainly incapable of satisfying this requirement. In other words, given the physical 'unreality' of threaded models, there may exist no energy function which is capable of uniquely recognizing the correct fold in all cases. One possible avenue for moving towards this goal may be to consider simulated folding pathways for each fold in the fold library, but for the time being, perfect fold recognition is but a distant dream.

The 'weak' fold-recognition problem is a far more practical formulation of the problem. Here the goal is to recognize and exclude folds which are not compatible with the given sequence with the eventual aim of arriving at a shortlist of possible conformations for the protein being modelled. At first sight this may not seem different from the goal of strong fold recognition, but the distinction is quite important. Even without a sophisticated fold-recognition method, weak recognition can be achieved by the application of simple common-sense rules. For example, if it is known that a protein is comprised entirely of alpha-helices (which might be known from circular dichroism spectroscopy, for example) then a large number of possible folds can be eliminated immediately (the correct fold could not be the all-beta immunoglobulin fold, for example). By applying a set of such rules, the 7000 or so possible folds could quickly be whittled down to a shortlist of say 10.

In reality, most, if not all of the published fold recognition methods really implement weak fold recognition. In the hands of an experienced user, however, who can make use of functional or structural clues in the prediction experiment, even weak fold recognition can be very powerful.

4.1 The domain problem in threading

Perhaps the main practical limitation of most 'weak' threading methods is that they are aimed at recognizing single globular protein domains, and perform very poorly when tried on proteins which comprise multiple domains. Unfortunately, threading cannot be used for identifying domain boundaries with any degree of confidence, and indeed the general problem of detecting domain boundaries from amino acid sequence remains an unsolved problem in structural biology. If the domain boundaries in the target sequence are already known, then of course the target sequence can be divided into domains before threading it, with each domain being threaded separately. Predictions can be attempted on very long multi-domain sequences, but in these cases the results will not be reliable unless it is clear that the matched protein has an identical domain structure to the target. For example, the periplasmic small-molecule binding proteins (e.g. leucine–isoleucine–valine binding protein) are two domain structures (two doubly wound parallel alpha/beta domains), but they all have identical domain organization. Proteins within this superfamily can thus be recognized by threading methods despite their multi-domain structure.

5 The future

One major difference between the academic challenge of protein structure prediction and the practical applications of such methods is that in the latter case there is an eventual end in sight. As more structures are solved, more target sequences will find matches in the available fold libraries—matched either by sequence comparison or threading methods. In terms of practical application, the protein-folding problem will thus begin to vanish. There will of course still be a need to better understand protein-folding for applications such as *de novo* protein design, and the problem of modelling membrane protein structure will probably remain unsolved for some time to come, but nonetheless, from a practical viewpoint, the problem will be effectively solved. How long until this point is reached? Given the variety of estimates for the number of naturally occurring protein folds, it is difficult to come to a definite conclusion, but taking an average of the published estimates for the number of naturally occurring protein folds and applying some intelligent guesswork, it seems likely that when threading fold libraries contain around 1500 different domain folds it will be possible to build useful models for almost every globular protein sequence in a given proteome. At the present rate at which protein structures are being solved, this point is possibly 15–20 years away. However, pilot projects are now underway to explore the possibility of crystallizing every globular protein in a typical bacterial proteome. If such projects get fully under way, which seems likely, then a complete domain fold library may be only five years away.

References

1. Jones, D. T., Taylor, W. R., and Thornton, J. M. (1992). A new approach to protein fold recognition. *Nature*, **358**, 86.
2. Godzik, A. and Skolnick, J. (1992). Sequence-structure matching in globular proteins: Application to supersecondary and tertiary structure determination. *Proc. Natl. Acad. Sci. USA*, **89**, 12098.

3. Ouzounis, C., Sander, C., Scharf, M., and Schneider, R. (1993). Prediction of protein structure by evaluation of sequence-structure fitness. Aligning sequences to contact profiles derived from three-dimensional structures. *J. Mol. Biol.*, **232**, 805.

4. Abagyan, R., Frishman, D., and Argos, P. (1994). Recognition of distantly related proteins through energy calculations. *Proteins: Struct. Funct. Genet.*, **19**, 132.

5. Overington, J., Donnelly, D., Johnson, M. S., Sali, A., and Blundell, T. L. (1992). Environment-specific amino-acid substitution tables—tertiary templates and prediction of protein folds. *Protein Sci.*, **1**, 216.

6. Matuso, Y., Nakamura, H., and Nishikawa, K. (1995). Detection of protein 3-D-1-D compatability characterised by the evaluation of side-chain packing and electrostatic interactions. *J. Biochem. (Japan)*, **118**, 137.

7. Madej, T., Gilbrat, J.-F., and Bryant, S. H. (1995). Threading a database of protein cores. *Proteins*, **23**, 356.

8. Lathrop, R. H. and Smith, T. F. (1996). Global optimum protein threading with gapped alignment and empirical pair score functions. *J. Mol. Biol.*, **255**, 641.

9. Taylor, W. R. (1997). Multiple sequence threading: An analysis of alignment quality and stability. *J. Mol. Biol.*, **269**, 902.

10. Bowie, J. U., Lüthy, R., and Eisenberg, D. (1991). A method to identify protein sequences that fold into a known three-dimensional structure. *Science*, **253**,164.

11. Taylor, W. R. and Orengo, C. A. (1989). Protein structure alignment. *J. Mol. Biol.*, **208**, 1.

12. Jones, D. T. (1998). THREADER : Protein Sequence Threading by Double Dynamic Programming. In *Computational methods in molecular biology* (ed. S. Salzberg, D. Searls, and S. Kasif). Elsevier, Amsterdam.

13. Sippl, M. J. (1990). Calculation of conformational ensembles from potentials of mean force. An approach to the knowledge-based prediction of local structures in globular proteins *J. Mol. Biol.*, **213**, 859.

14. Rost, B. (1997). Protein fold recognition by prediction-based threading. *J. Mol. Biol.*, **270**, 1.

15. Jones, D. T. (1999). GenTHREADER: An efficient and reliable protein fold recognition method for genomic sequences. *J. Mol. Biol.*, **287**, 797.

16. Murzin, A. G. and Bateman, A. (1997). Distant homology recognition using structural classification of proteins. *Proteins Suppl.*, **1**, 105.

17. Russell, R. B. and Barton, G. J. (1994) Structural features can be unconserved in proteins with similar folds—an analysis of side-chain to side-chain contacts, secondary structure and accessibility. *J. Mol. Biol.*, **244**, 332.

18. Orengo, C. A., Jones, D. T., and Thornton, J. M. (1994). Protein superfamilies and domain superfolds. *Nature*, **372**, 631.

19. Kraulis, P. J. (1991). Molscript—a program to produce both detailed and schematic plots of protein structures. *J. Appl. Crystallogr.*, **24**, 946.

20 Jones, D. T., and Thornton, J. M. (1996). Potential energy functions for threading. *Curr. Opin. Struct. Biol.*, **6**, 210.

Chapter 2

Comparison of protein three-dimensional structures

Mark S. Johnson and Jukka V. Lehtonen

Department of Biochemistry and Pharmacy, Åbo Akademi University, Tykistökatu 6 A, 20520 Turku, Finland.

1 Introduction

In this chapter we define the different types of questions that may be asked through the comparison of the three-dimensional (3-D) structures of proteins, how to make the comparisons necessary to answer each question, and how to interpret them. We shall focus on the different strategies used, and the assumptions made within typical computer programs that are available.

Protein structure comparisons are often used to highlight the similarities and differences among related—*homologous*—3-D structures. Homologous proteins are descended from a common ancestral protein, but have subsequently duplicated, evolved along separate paths, and thus changed over time. The independent evolution of related proteins with the same function, *orthologous* proteins, which are found in different species, and the *paralogous* proteins, which have evolved different functions, all retain information on the original relationship. The amino acid sequences change over time reflecting the mutations, insertions and deletions that occur in their genes during evolution, and for many proteins the sequences themselves are so similar that common ancestry is apparent. For others, the sequences can be so dissimilar that the case for *homology* may be difficult to make on the basis of the primary structure. Nonetheless, comparing the 3-D structures when they are available can identify homologous proteins. This is possible since the evolution of proteins occurs such that their folds are highly conserved even though the sequences that encode them may not be recognizably similar.

Homologous proteins are often compared in order to highlight features (typically the amino acids and their relative orientations to one another), which have come under strong evolutionary pressure not to change because of structural and functional restraints placed on them. Conversely, differences in an otherwise conserved active site or binding site are used to explain differences in observed function.

Dayhoff and coworkers (1) long ago predicted that about 1000 different protein families should exist in nature, and it has become clear over recent years that most newly-solved 3-D structures do fall into an existing family of structures (2, 3). The approximately 100 000 proteins encoded in the human genome, whose sequences will be known early in this century, will fall within this limited number of families. Thus, one key bioinformational goal has been to compare and classify all proteins and their component domains into family groups, and one immediate goal is to solve at least one representative structure for each sequence family that is not obviously connected to any existing structural family. This single representative structure can then be used in knowledge-based modelling (4) to estimate the 3-D structures for other members of the family.

Comparisons are also made among non-homologous proteins to try and highlight structural features that are locally similar, but whose present-day sequences have not arisen as a consequence of evolutionary divergence from a common ancestor. Classic examples include the active site similarities among serine proteinases, subtilisins and serine carboxypeptidase II (5), each of which invoke the participation of histidine, serine and an aspartic acid in their proteolytic mechanism of action. The folds are different and the relative positions of these key amino acids along the sequence are different too. In the 3-D structures, however, the residues are similarly positioned to reproduce a common catalytic mechanism that has been exploited by nature on at least three separate occasions. Comparisons among non-homologous proteins can highlight structural units that are common features of the protein fold and comparisons have been made to classify amino acid conformations, regular elements of secondary structure (helices, strands, turns), supersecondary structure, and cofactor and ligand binding sites.

The comparison of protein structures can be achieved in many different ways. In this chapter, we present several of the basic procedures used in the wide variety of programs that have been developed over the years. These methods range from rigid-body comparisons, to methods more typical of sequence comparisons—dynamic programming, and to those methods that employ Monte Carlo simulations, simulated annealing and genetic algorithms to find solutions for combinatorially-complex structural comparison problems. We will describe methods that demand partial solutions as input to the procedure, as well as strategies for automatic hands-off solutions; and approaches to both homologous and non-homologous structural comparisons.

2 The comparison of protein structures

2.1 General considerations

The optimal superposition of two identical 3-D objects can be determined exactly. This only requires the calculation of (a) a translation vector to place one copy of the object over the other at the origin of the co-ordinate system and (b) a rotation matrix that describes the rotations needed to exactly match the two copies of the object. The *translation vector* describes movements along the x, y, and z directions in the co-ordinate system. The *rotation matrix* describes the α, β, and γ rotations in the three orthogonal planes. One of the main tasks of many super-positioning procedures is to define these values and then to apply them to the co-ordinates of the objects and they will then be superposed on each other. Two identical objects will have all points superposed exactly.

The major difficulty with non-identical objects, such as a pair of protein structures, is that they typically have different numbers of amino acids, different amino acids with different numbers, types and connectivities of atoms. Furthermore, amino acids present in one structure can be missing in the other: insertions and deletions—the gaps seen in a sequence alignment. Thus, except in the case of one protein co-ordinate set being compared with itself, no two proteins will have atoms in exactly identical positions. A protein whose structure has been solved several times will also vary with overall differences in the main chain co-ordinates of no more than about 0.3 Å, but they will be different.

The superposition of most protein structures as rigid-bodies, therefore, is not straightforward, and several different considerations need to be resolved in advance of the comparison. These include:

(a) Which atoms will be compared between the molecules?

(b) How will the dissimilarity or similarity between relative positions of matched atoms be taken into account?

(c) Should the structures be compared as rigid-bodies (in most cases, resulting in a partial alignment of the most similar regions, which can be displayed graphically)? Or have significant structural shifts occurred that require a procedure that can accommodate these changes (typically providing the complete alignment of the sequences, including gap regions, on the basis of the structural features compared)?

(d) How will one define what constitutes an equivalent matched set of co-ordinates between non-identical objects where exact matching of atoms will only rarely be seen?

(e) How will the program be initially seeded? Many methods need to be supplied with co-ordinates of a set of equivalent atoms at the onset of the comparison, a minimum of three matched pairs, thus requiring some information on the likely superposition of the two structures in advance of comparison.

(f) How will the quality of the structural comparison that results be assessed?

2.2 What atoms/features of protein structure to compare?

Depending on what question you wish to answer by comparing a pair of structures, the choice of which atoms' co-ordinates will be superposed can be crucial. For example, to look at similarities/differences surrounding a bound cofactor common to two proteins, you may choose to superpose all or some of the atoms of the cofactor, apply the translation vector and rotation matrix to the entire co-ordinate file—protein and cofactor included. Alternatively, the backbone co-ordinates of the proteins could be superposed and the relative positions of the cofactors examined after the superposition.

It is usually not very useful to compare atoms of amino acid side chains when making global structural comparisons. Different amino acid types have different number of atoms and different connectivities that can preclude their direct comparison. Residues, even identical ones, will have different conformations, especially when they are located at the solvent-exposed surface of the proteins. However, there are situations where the local comparison of side chains can be very useful, for example, in the comparison of residues lining an active or binding site especially when different ligands are bound to the same or similar structures.

For most general methods, which aim to superimpose two proteins over the maximum number of residues, the Cα-atom co-ordinates are typically employed (all atoms of the protein backbone and even the side chain Cβ-atom, but excluding the more positionally-variable carbonyl O, can also be used). (Except where noted, we will consider Cα-atom co-ordinates in the protocols described herein.) Whereas the side chain conformations can vary wildly between matched positions in two structures, the Cα-atom or backbone trace of the fold is typically well conserved, with regular elements of secondary structure, the α-helices and

Table 1 Examples of features[a] of proteins that can be used in comparisons

Properties

(a) Residues	(b) Segments
Identity	Secondary structure type
Physical properties	Amphipathicity
Local conformation	Improper dihedral angle
Distance from gravity centre	Distance from gravity centre
Number of neighbours in vicinity	Average C^α density
Position in space	Position in space
Global direction in space	Global direction
Main chain accessibility	Main chain accessibility
Side chain accessibility	Side chain accessibility
Main chain orientation	Orientation relative to gravity centre
Side chain orientation	
Main chain dihedral angles	

Relations

(a)	(b)
Disulfide bond	Relative orientation of two or more segments
Vectors[b] to one or more nearest neighbours	Vectors[b] to one or more nearest neighbours
Distances to one or more nearest neighbours (e.g. atom pairs or contact maps)	Distances to one or more nearest neighbours
Change in number of neighbours in vicinity	
Ionic bond	
Hydrogen bond	
Hydrophobic cluster	

[a] See refs 7, 8, 10, 11.

[b] Vector defines both distance and direction in the local reference frame.

β-strands, matching closely and sequentially along the fold of the two structures. Differences in Cα-atom traces are more often seen at loop regions that connect the strands and helices in proteins: Frequently these loop regions are exposed to the solvent at the surface of the protein and thus have fewer constraints placed on their conformations.

For more dissimilar protein structures, rigid body movements and other structural changes can occur in one structure relative to the other. When this happens, rigid-body comparisons of the 3-D structures can often lead to poorly matched structures, although the folds are the same. If these changes are not large, then dynamic programming procedures (6) that consider only Cα-Cα atom distances or other structural properties of the amino acids (*Table 1*) after an initial rigid-body comparison can be quite effective in matching all residues from the protein structures (7–9). Others have described automated procedures that involve the comparison of structural relationships that require special techniques to solve these problems of combinatorial complexity (7, 8, 10, 11).

Protocol 1

Features used for the comparison of protein 3-D structures

Distances between atomic co-ordinates are often used (a) for more similar proteins where rigid-body shifts of one structure relative to another are not a significant factor, (b) to illustrate the degree of structural change, or (c) where a local comparison of a site of interest—active site or binding site—is desired for visualization purposes. Where significant changes to the structures have occurred, other structural features, which are not as sensitive to these relative structural shifts, can be compared in addition to atomic co-ordinates.

Rigid-body structural comparisons

1   Choose the atoms for comparison that are appropriate for the question to be asked. Most often, but not necessarily, the Cα-atom co-ordinates are used by default.

2   Comparisons will then be based on the distances between atoms that are considered to be equivalent. For rigid-body methods, a distance cut-off is used to define equivalent matched positions. Typically, the cut-off value is on the order of 3 Å, although values between 2.5 Å and 4.5 Å have been used. Lower values are more restrictive and will lead to fewer aligned positions in more dissimilar structures.

Structural feature comparisons

1   Features of individual atoms, residues or segments of residues, both properties of and relationships between individual atoms, residues or segments, are considered either separately or in combination with each other as a basis for structural comparisons (see *Table 1* and refs 7, 8).

2   Comparisons will be based on differences/similarities between potential matched regions in the two structures in terms of the features compared. An alignment

Protocol 1 continued

algorithm is used to give the best 'sequence alignment' based on the structural features that have been supplied.

(a) Property comparisons may require an initial alignment (e.g. rigid-body).

(b) Relationships can be aligned by a variety of methods, e.g. Monte Carlo simulations (11), simulated annealing (7), double dynamic programming (10), genetic algorithms.

3 The structures can subsequently be superposed according to the matches in the alignment, but a single global superposition may be meaningless when large movements, such as domain movements, have taken place. In that case, each domain should be superposed separately.

Dynamic programming methods can align structures on the basis of differences/similarities between any number and combination of *properties*—which are features of individual residues or segments of residues contiguous in sequence. In order to compute the difference or similarity between positions in a structure, for example on the basis of Cα–Cα distances, an alignment is required to give an estimate of the distances between atoms in the structures. Other structural properties, such as residue solvent accessibility, can be used with dynamic programming directly, but may provide less useful information for the comparison. *Relationships*—features of multiple non-sequential residues (*Table 1*): e.g. patterns of hydrogen bonding, hydrophobic clusters, Cα-atom contact maps—can also be compared. Monte Carlo simulations (11), simulated annealing (7), and double dynamic programming (10) have all been used to equivalence relationships among residue sets from structures. Each of these methods gives an alignment of the structures in the form of a sequence alignment, but to visualize the results of the comparison, a rigid-body comparison would still be required. This could be made over all matched positions or over those positions that matched 'best' according to the comparison criteria used. The global rigid-body superposition based on the alignment may also be unsatisfactory if large structural changes have taken place. To accommodate very large changes, such as domain movements, the domains can be superposed separately.

2.3 Standard methods for finding the translation vector and rotation matrix

For methods that compare the relative atomic positions in two structures, A and B, and produce the superposed co-ordinates as output, it is necessary to determine a translation vector and the rotation matrix that, when applied to the original co-ordinates, will generate the new co-ordinates for the superposed proteins. Firstly, the centre-of-mass of the each protein is translated to the origin of the co-ordinate system. Secondly, one of the structures is rotated about the three orthogonal axes in order to achieve the optimal superposition upon the other structure. Because the atoms chosen for comparison will not match exactly in terms of their relative atomic positions after superposition, a least-squares method is typically used to achieve the optimal superposition.

The following function minimizes the residual δ, which is expressed mathematically as:

$$\delta = \sum_{i=1}^{N} w_i \, (\overset{\leftrightarrow}{A}_i^{eq} - \Re \overset{\leftrightarrow}{B}_i^{eq})^2$$

where $\Re$ is the rotation matrix being sought that minimizes the differences between a total of N equivalent co-ordinate sets $\overset{\leftrightarrow}{A}^{eq}$ from the first protein and $\overset{\leftrightarrow}{B}^{eq}$ from the second protein; w_i is a weighting that can be applied to each ith pair of equivalent positions.

Numerous methods have been developed to solve this pairwise least-squares problem in a variety of different ways (12–16). Others have described more general methods suitable for the least-squares comparison of more than two three-dimensional structures (17, 18) In our experience, the method of Kearsley (19) is a straightforward and simple means to obtain the optimal rotation matrix for a set of equivalent co-ordinates. We will only consider this procedure here (*Protocol 2*).

The major obstacle to solving the least-squares problem is that matched atom pairs from the two structures to be compared need to be specified to the algorithm at the beginning of any calculations. Thus, the computer program requires some idea of the final alignment before it can proceed. There are common situations where the comparisons would be made over a pre-defined set of residues: for example, (a) comparisons over residues that line an active site or binding site—to highlight similarities and differences over those positions; (b) comparisons of independent structure solutions for the same protein. In these cases, the atomic positions to be compared are usually known *a priori*, and a single round of rigid-body comparison is sufficient to obtain the optimal match. Frequently, however, global comparisons are made between proteins where the best-matched positions are not obvious in advance. In the case of similar protein structures, the requirement of an initial set of matches to seed the comparison is inconvenient at best, requires the pre-analysis of the proteins involved, and in the case of more dissimilar proteins, may be difficult to define. Additionally, we have often observed that when part of the answer is specified at the beginning of the comparison, then the final solution can be prejudiced to give a final result that is not necessarily the optimal one: The comparison was locked into a set of possible solutions by the information supplied to seed the procedure. Despite these criticisms, there are many good methods that employ this strategy.

For example, Sutcliffe *et al.* (16) specify a set of at least 3 Cα-atoms common to the two structures (3 positions define a unique plane in each structure). Good candidates for these common residues, supplied *a priori*, can be conserved residues at an active site or ligand binding site, be positions conserved in terms of the sequence similarity, or can be equivalent positions observed to form part of the common fold when the proteins are examined on a graphics device. This and other similar methods use an iterative procedure to progress towards better

21

and better solutions that incorporate more and more equivalent atom pairs. (Later in this chapter we will detail several automatic strategies that have been used to get around this need for predetermining a set of equivalent atoms at the onset of the structural comparison.)

In the equation describing the residual (above), $\overset{\omega}{A}^{eq}$ and $\overset{\omega}{B}^{eq}$ contain the x, y, and z axes co-ordinates for exactly the same number of atoms from each of the two structures. These atoms are termed *equivalent* positions, and are those aligned positions that the superposition will now be calculated for. All other atoms in the molecules are ignored in determining the superposition, but the translation vector and the rotation matrix determined on the basis of these equivalent positions is subsequently applied to all atoms in the co-ordinate file, including any bound ligand, metal ions, and water molecules. Here, we will detail how to calculate the translation vector for each protein and describe one simple yet elegant method for determining the rotation matrix, developed by Kearsley (19, 20), which we use as the method of choice for our own procedures (*Protocol 2*).

Protocol 2

Rigid-body structural comparisons: translations and rotations

This protocol details the steps required to optimally superpose the equivalent atom co-ordinates from two proteins.

Data required

The co-ordinates of all atoms in the proteins' co-ordinate file (minimum of the Cα-atom co-ordinates) and the matched equivalent atoms in the two proteins.

The translation vector

1 Calculate the centre of mass from the x, y, and z co-ordinates for each set of equivalent atoms $\overset{\omega}{A}^{eq}$ and $\overset{\omega}{B}^{eq}$ from the two structures. For N atoms in the equivalent set of the first protein:

$$\overset{\overline{\omega}}{T}_A = \sum_{i=1}^{N} \overset{\omega}{A}^{eq}_i / N$$

In other words, sum all of the x co-ordinates together and divide by N to give the average x co-ordinate for the equivalent set of atoms; repeat for the y and z co-ordinates. Repeat for the corresponding N equivalent atoms in the second structure:

$$\overset{\overline{\omega}}{T}_B = \sum_{i=1}^{N} \overset{\omega}{B}^{eq}_i / N$$

Thus, the centre of mass is a single x, y, and z co-ordinate set for each of the proteins.

2 Translate both structures, all atoms in the file, so that their centres of mass (according to the set of equivalent atoms used) are located at the origin of the co-ordinate system. For *every* atom i in the first structure:

$$\overset{\omega}{A}^{all}_i (trans.) = \overset{\omega}{A}^{all}_i (old) - \overset{\overline{\omega}}{T}_A.$$

In other words, subtract the x, y, and z co-ordinate values for the centre of mass from the x, y, and z co-ordinate values for every atom in the co-ordinate file. Repeat for the second structure:

$$\overset{\omega}{B}{}^{all}_i\,(trans.) = \overset{\omega}{B}{}^{all}_i\,(old) - \overset{\overline{\omega}}{T}_B.$$

The rotation matrix: the Kearsley method (ref. 19) minimizes the average difference between sets of atoms using quaternion algebra

1. Generate a symmetric 4×4 matrix by adding selected combinations of differences and sums of co-ordinates calculated for each matched pair of equivalent atoms to the elements of the matrix (19). These are the co-centred co-ordinates, but only the co-ordinates of equivalent matched atom pairs, $\overset{\omega}{A}{}^{eq}_i$ and $\overset{\omega}{B}{}^{eq}_i$ are used at this stage.

2. Diagonalize the 4×4 matrix in order to obtain its eigenvalues and eigenvectors (see ref. 21 for general procedures).

3. Select the lowest eigenvalue and use elements of the corresponding eigenvector to construct the 3×3 rotation matrix $\mathfrak{R}$ (see ref. 19 for details).

4. Multiplication of each co-ordinate in the second structure B by $\mathfrak{R}$ will produce the superposition of the entire structure onto protein A, where the average distance between matched atoms of the equivalent set is a minimum: $\overset{\omega}{B}{}^{all}_i\,(trans.,rot.) = \mathfrak{R} \times \overset{\omega}{B}{}^{all}_i\,(trans.)$.

5. The selected eigenvalue divided by the number of atom pairs in the equivalent set is equal to the square of the RMSD after rotation. $\mathfrak{R}$, calculated above, leads to the superposition whose RMSD is a minimum for these sets of equivalent atoms.

2.3.1 Structural alignment of sequences

In *Figure 1*, is shown the loss of superposed Cα-atoms in globin comparisons as the percentage sequence identity decreases. As an alternative to rigid-body structural comparisons, especially when the rigid-body structural similarity is reduced due to modest structural alterations, other methods have been developed that provide the alignment of the sequences of the structures. Nonetheless, rigid-body comparisons are often used in combination with these other procedures or for visualisation of the results.

For example, the dynamic programming algorithm described below can make comparisons on the basis of Cα–Cα atom distances, as well as other features (see *Table 1*).

(a) As we have stated above, a rigid-body comparison is often needed in order to make comparisons of structural properties suitable for dynamic programming alignment.

(b) The dynamic programming method is often used in conjunction with rigid-body super-positioning methods in order to efficiently assign equivalent matches.

(a) 52% id, 140 equivalences, RMSD = 0.82Å   (c) 35% id, 115 equivalences, RMSD = 1.46Å

(b) 43% id, 136 equivalences, RMSD = 1.32Å   (d) 16% id, 97 equivalences, RMSD = 1.72Å

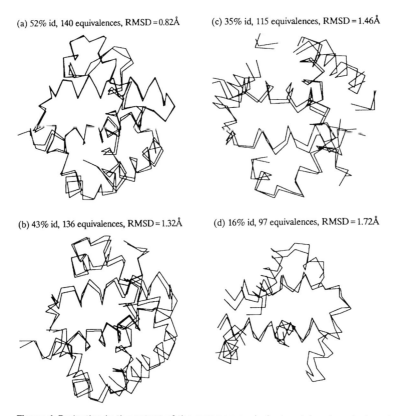

Figure 1 Reduction in the extent of the common equivalent matches in pairwise structural superpositions as a function of decreasing percentage sequence identity. Traces of the backbones are shown for Cα-positions within 2.5 Å after rigid-body superposition with the computer program MNYFIT (16). The haemoglobin α-chain of *Pagothenia bernacchii* (Protein Data Bank (PDB, ref. 51) code: 1PBX) is aligned in (a) with the α-chain of equine haemoglobin (2MHB) and in (b) with the β-chain of human haemoglobin (2HHB). (c) The human haemoglobin β-chain (2HHB) aligned with the sea lamprey globin (2LHB). (d) The erythrocruorin of *Chironomous thummi thummi* (1ECD) aligned with the leghaemoglobin of *Lupinus luteum* (1LH1). (From ref. 4, with permission.)

(c) The dynamic programming algorithm produces a full alignment of all positions in the structures (residues are aligned with each other or with gaps), while the rigid-body methods align fewer and fewer potions in the structures as the sequence similarity decreases (*Figure 1*).

(d) Dynamic programming algorithms do not give a superposition of the structures suitable for visualization. This can be obtained from the alignment by applying the rigid body method to the defined matched pairs.

(e) Dynamic programming can often lead to alignments of the structures where rigid-body movements have occurred in the structures themselves. For example, the large movements of the entire domains seen in the liganded and unliganded structures of the periplasmic bacterial lysine–arginine–ornithine binding protein (*Figure 2*). Rigid-body comparisons can be applied,

24

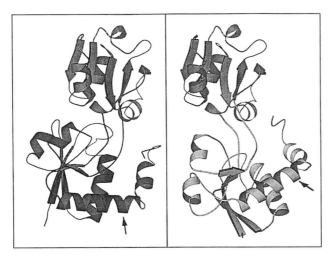

Figure 2 Two different conformations of the 3-D structure for the same protein, the lysine–arginine–ornithine binding protein from *Salmonella typhimurium*. *Left*: the structure of the protein in complex with lysine (1LST), lysine not shown. *Right*: the uncomplexed structure (2LAO). The smaller domain on the upper part of the figures is in same orientation and the arrow pointing to the Cα-atom of Glu 216 illustrates the magnitude of the movement of the larger domain at the bottom of the figure. Figure prepared with MOLSCRIPT (52).

however, to the domains separately to pinpoint any changes within each domain that have occurred upon ligand binding; while superpositioning on one domain can be used to highlight the relative movements that have occurred between the domains upon binding.

2.4 Standard methods to determine equivalent matched atoms between structures

There is no exact definition of topological equivalence, and the criteria used can vary from method to method. In rigid-body superposition methods, a distance cut-off between equivalent atoms is frequently used. In methods were other structural features are considered, all aligned positions might be considered to be topologically equivalent between two structures, or they may be assigned according to the degree of positional similarity of features used to make the comparison.

2.4.1 Definitions of structural equivalence: the alignment

In determining a set of equivalent atom sets, distance criteria are often used. After one structure has been superposed on another, topological equivalent atoms can be limited to those atom types under consideration that are within a distance cut-off value. The Euclidean distance, D, between two points is:

$$D_i = \sqrt{\left(x_{A(i,eq)} - x_{B(i,eq)}\right)^2 + \left(y_{A(i,eq)} - y_{B(i,eq)}\right)^2 + \left(z_{A(i,eq)} - z_{B(i,eq)}\right)^2}.$$

In rigid body comparisons, where the Cα-atoms of the protein backbone have been used as a basis for comparison, a distance cut-off typically in the range 2.5 Å to 4.5 Å has been used. Values above 3 Å lead to more multiple matches to a single atom: the distance between two consecutive Cα-atoms along the protein backbone is around 3.5 Å. Lower values will reduce the number of equivalent matches when more dissimilar proteins are compared. Distances or dissimilarity measures will also be required for the comparison of other structural features, both properties and relationships, see ref. 7 for example. Common to both rigid-body methods, which rely on simple distance data, and other methods, which incorporate other types of information into the alignment process, is the need to determine the matching of locations between the structures to be superposed (*Protocol 3*). This can be part of an iterative procedure to provide a new set of equivalent atoms that are then used to determine a new translation vector and rotation matrix in order to improve a match. This is also one of the final steps in any comparison procedure, where the resultant alignment is determined. Three basic approaches have been used: (a) dynamic programming, (b) graph theoretical match list handling and clique detection methods, and (c) methods more suitable for solving combinatorially-complex matching problems.

The Needleman and Wunsch (6) method is a convenient fast method for aligning proteins. By scoring all possible pairs of matches between two structures, the method insures that the optimal scoring solution is found for the scoring scheme employed. The method accommodates a loss of elements in one structure relative to another—the gaps corresponding to insertions and deletions. Thus, the method provides a full alignment where every residue position in each protein is matched to either a residue position in the other protein or a gap. Thus, this method can efficiently resolve the multiple matching and many combinatorial problems seen with the list sorting procedure. Once structural relationships have been equivalenced between a pair of structures, this information can also be used within the dynamic programming method.

With the match list sorting procedure, for example ref. 22, possible equivalent matches between the proteins are tabulated: matches of protein B to each position in protein A in one list, and matches of protein A to protein B in a second list. These lists contain both authentic matches of conserved structure, chance matches that need to be eliminated from the lists and multiple matches between one element in one protein to several different elements in the other protein. The challenge, then, is to cull these lists by keeping the best matches (i.e. matches that can extend a series of previous matches, have a good matching score or give a good fit), removing structurally unlikely matches (matches that are not co-linear—are out of sequence with other matches—and isolated matches that do not extend further other matches), and by reducing multiple matches to single matches.

A more elaborate approach was introduced by Mitchell *et al.* (23). Their method does not filter out extraneous matches, but instead tests each combination of matches to find the optimal equivalent set. As a result, a 'clique', the maximum sub-graph common to two graphs representing the structures is found. The clique detection algorithm is based on graph theory and offers a way to find similar parts of structures that have not been superimposed. The basic idea is to represent each structure as a graph of nodes and vertices. Each node corresponds to either an atom, piece of main chain, secondary structure element, or similar definite piece of structure. Each vertex is a relation between two nodes in a structure: the distance between the atoms, vector from one atom to another (both distance and direction in some co-ordinate frame), distance and angle between two secondary structure elements, or more a more complicated distance measure involving other properties of the nodes. If two structures contain a similar substructure, then the nodes belonging to that substructure are connected in both structures by very similar vertices. The task is to find the maximal common sub-graph from the set of possible common sub-graphs. While this is a NP-complete task, it is feasible due (a) efficient search algorithms evolved within graph theory, and (b) the use of (few) secondary structure elements (SSEs) as the compared pieces of the structures instead of (many) atoms. Several other programs have been described that use a very similar approach (see ref. 24 and citations therein); the main differences are in the ways structures are represented and in the method used to reduce the search space for efficiency.

The comparison of relationships among features in one structure relative to another is a powerful addition to any structural comparison procedure (see ref. 7 for an excellent discussion). Relationships, such as patterns of hydrogen bonding, involve the comparison of a minimum of two residue positions for every hydrogen bond in both structures. In certain cases, e.g. in the method of Taylor and Orengo (10), relationships—in this case inter-atomic vectors, are compared using their novel double dynamic programming method. More often, the matching of relationships is treated as a combinatorially-intensive task. There are lots of candidate pairs of hydrogen bonds in each structure and matching them relies on methods such as simulated annealing, Monte Carlo simulations and genetic algorithms.

Protocol 3

The alignment: determination of equivalent pairs

Methods used to find the optimal match between entire structures or between parts of structures consisting of the best matching regions of the structures. Equivalent or matched positions are defined by the user (i.e. property distances within a cut-off value) or by the strategy of the method employed (e.g. all matched positions produced by dynamic programming methods).

Dynamic programming methods (6)

1 Construct a matrix with dimensions equivalent to the lengths of the structures to be compared.

2 Each cell in the matrix corresponds to a residue in the first protein matching a residue in the second protein. The matrix accommodates all possible alignments.

3 Cells are filled in with a score relating each two matched positions. These scores may be distances between $C\alpha$-atoms, for example, distance scores based on other

Protocol 3 continued

features of the protein (see Table 1), or similarity scores derived from distances. In this description, we will refer to a matrix filled with similarity scores derived from distances.

4 Beginning at one corner (amino-terminal end or carboxy-terminal end of the sequences) of the matrix and heading towards the opposite corner, sum diagonal values to the current position if they are the best score (a residue–residue match), or sum with an off-diagonal score minus a penalty (indicates a possible gap in one protein or the other).

5 The largest value found at one edge of the matrix specifies the first two aligned positions and gives the optimal alignment score for the comparison.

6 The full alignment that produced the optimal score can be traced beginning at the highest value and progressing towards the opposite side of the matrix by following the next best score in the matrix. When the next highest value is on the diagonal, residues are matched in sequence; when an off diagonal score (less a penalty) is the next best choice, then a gap is indicated.

7 This method produces the full alignment including gap regions, but elements within a cut-off value can be used to determine the rigid-body superposition of the structures.

Clique detection methods (23, 25)

1 Represent each structure as a graph of nodes ($C\alpha$-atoms or secondary structure elements) and vertices connecting the nodes. Each vertex is a distance between the connected two nodes (atoms).

2 List for each vertex in structure A all such vertices in structure B, which are similar within an error threshold (i.e. vertices connecting the same kind of nodes with similar distances).

3 Find the maximal common sub-graph (largest set of nodes and vertices, which exists in both structure graphs) using a tree search algorithm, Monte Carlo simulation, or a genetic algorithm. Each vertex in the common sub-graph corresponds uniquely to one vertex in both structures A and B.

4 The nodes included in the sub-graph are equivalent for the two structures. If the nodes are atoms, the superposition can be made directly (see Protocol 2). Also, the secondary structure elements can be superimposed as if they were atoms of a rigid molecule, or the $C\alpha$-atoms within the SSEs can be superimposed.

Match list approaches (22)

This method is a variation of the clique detection method, which assumes that the structures are initially superimposed, but equivalent matches are not known.

1 In the case of $C\alpha$-$C\alpha$ distance comparisons, create two lists, one for each protein A and B.

   (a) In one list, tabulate all $C\alpha$-atoms in protein B with matches within a cut-off distance, say 3.5 Å, to a position in protein A.

    (b) In a second list, tabulate all Cα-atoms in protein A with matches within a cut-off distance to a position in protein B.

2   Filter from the list the poorest matches to reduce the number of matches to a unique set of equivalent matches:

    (a) Remove matches that are not part of a contiguous run of at least 4 Cα-atoms.

    (b) Reduce multiple matches from one protein to a single Cα-atom in the other protein, e.g. does one of the matches extend a contiguous run of existing matches?

    (c) If there are still multiple matches remaining, then the match with shortest distance is kept and the others are removed.

Comparisons of relationships (7, 10, 11, 21)

1   The matching of relationships among features of one structure with relationships among features of another structure is accomplished using one of several different techniques.

    (a) Monte Carlo simulations (11, 21).

    (b) Simulated annealing (7, 21).

    (c) Double dynamic programming (10)

    (d) Genetic algorithms (22, 26, 27) can also be used.

2   The matched relationships may be insufficient in themselves to accurately align the 3-D structures, and thus would be combined with the feature comparisons within a dynamic programming procedure, for example, to give the final alignment (7).

2.5 Quality and extent of structural matches

Once a structural alignment has been made, a score or scores can be assigned to the alignment that give an indication of the quality and the extent of matching between the two structures. With methods that iteratively improve a structural comparison, an evaluation score is necessary to monitor the improvement at each cycle of comparison, and to indicate when the program should stop because no further improvement in the alignment could be obtained. The final alignment scores can be used to compare different protein comparisons within a family and provide useful indications of the phyletic ancestry of the proteins (e.g. 8, 28). Among the most frequently used key indicators of the 'goodness' of a structural comparison include the root mean squared deviation (RMSD), the number of topologically-equivalent atoms matched in the comparison, and the alignment score that is obtained.

2.5.1 Root mean squared deviations

The RMSD is commonly used to indicate the goodness of fit between two sets of co-ordinates. Often, but not always, the RMSD value is quoted for only those matched Cα-atoms that are within a specified distance cut-off, say atoms within 3.0 Å of each other after the proteins have been superposed. In this case and given the cut-off value of 3 Å, the RMSD obtained and each of the $C\alpha$–$C\alpha$ atom distances contributing to the RMSD will be less than the 3 Å. Alternatively, the RMSD can be calculated over all matched $C\alpha$–atom pairs, regardless of the distance between the superposed atoms. Of course, the RMSD can also be calculated between sets of any type of superposed atoms, not just $C\alpha$–atom pairs as illustrated in *Protocol 4*.

Protocol 4

Root mean squared deviations (RMSD)

The RMSD gives a measure of the average level of deviations over the matched atoms that are included in the calculation. Given the same number of equivalent atom pairs, a smaller value indicates a better superposition than does a larger value.

Data required

- Co-ordinates of the equivalent sets $\overset{\omega}{A}_i^{eq}$ (trans.), $\overset{\omega}{B}_i^{eq}$ (trans.,rot.).

Method

1   Calculate the Euclidean distance between each pair of equivalent atoms $\overset{\omega}{A}_i^{eq}$ (trans.) and $\overset{\omega}{B}_i^{eq}$ (trans.,rot.).

2   Take the sum of all squared distances D, and divide by the number of pairs, N, to give the mean.

3   Calculate the square root of the mean squared distance to obtain the RMSD.

4   Thus, the

$$\text{RMSD} = \sqrt{\sum_{i=1}^{N} D_i^2 / N}$$

2.5.2 Topological equivalent atoms pairs

Another criterion that is used to gauge the extent or quality of a superposition of two structures is the number of atom pairs that superpose within a distance cut-off. Structure comparison methods usually try to maximize the number of superposed equivalent atoms while minimizing the RMSD over those equivalent atoms.

Note that two different sets of superposed structures, given the same cut-off value, can have the same number of equivalent matches, but with different RMSD values over those matches. The match with the lower RMSD would be considered the more similar pair. Conversely, one structural comparison, for example, may produce 121 matches with an RMSD of 2.1 Å, while a second comparison matches 50 atom pairs with an RMSD of 1.2 Å: the comparison with the 121 matches would be considered the better match.

2.5.3 Structural alignment scores

For structural alignment methods that employ dynamic programming in order to produce a complete alignment of the structures, including gaps, a key measure of the alignment quality is the alignment score corresponding to the overall optimal structural superposition. This value includes scores for matching all positions and penalties for every gap that appears in the alignment. The alignment score is composed of the values placed into the matching matrix during the dynamic programming procedure. In the case of Cα-atom based comparisons, the residue–residue matching scores would be the distances between the atoms. In the case of procedures that consider other criteria, e.g. those features listed in *Table 1*, the alignment score would include the scores attributed to matches of residue positions according to those features.

The raw alignment score is useful during iterative procedures to provide an indication of the progress of the superposition. Within a family of homologous 3-D structures, the alignment score, normalized for the length of the smaller protein or for the number of matched residues along the sequences, can be compared to give an idea of the mutual structural relationships among the family members.

3 The comparison of identical proteins

3.1 Why compare identical proteins?

The simplest type of comparison of 3-D structures involves the comparison of two (or more) sets of co-ordinates for the same protein. Self-comparisons are often used to reveal:

(a) Similarities/differences between independent solutions of crystal structures.

(b) Similarities/differences among sets of structures, generated using distance geometry, and consistent with distance information obtained in NMR spectroscopy.

(c) Similarities/differences between structures obtained using X-ray diffraction and NMR spectroscopy.

(d) Similarities/differences that occur between apo- and holo-protein structures: alterations in structure that occur upon binding ligands, cofactors, metal ions, etc.

(e) Similarities/differences of two structures after superposing on an identical ligand or subset of residues or co-ordinate positions.

3.2 Comparisons

In the comparison of identical proteins that have 3-D structures that differ to varying degrees, it is needed to compare the structures using a rigid-body approach one time only (*Protocol 5*). No iteration is necessarily required to achieve the best result, since one would typically supply all atom positions in the structure for comparison. Likewise, no pre-comparison is necessary to supply a seed set of residues for the comparison. In practice, iterative procedures are used. Again, if big differences in the structures are anticipated, e.g. the relative domain movements in *Figure 2*, then this approach may not be appropriate except to provide an RMSD value that is an indication of the relative changes to the structures.

Protocol 5

Similarities among different structures of identical proteins

Finds regions of high structural similarity between different solutions of the structure of the same protein.

Data required

- Co-ordinates, minimum of Cα-atoms, for all structures

Method

1　No alignment between the proteins is necessary[a] since the proteins are identical and each position maps 1:1 in sequence along the protein.

2　A single application of a comparison algorithm (see Protocol 2) is sufficient to obtain the optimal result over all of the compared atoms.

3　Calculate the RMSD over all atoms or those within the cut-off distance, as desired (see Protocol 4).

4　Iterative methods (see *below*), seeded by some key positions, can be used also.

5　By adjusting the cut-off value used to define equivalent matched atoms to lower values, the most similar structural regions may be identified and hence, the differences pinpointed too.

[a] Note that different data sets from different sources do not necessarily contain the same amino acids or atoms for the same protein.

4 The comparison of homologous structures: example methods

4.1 Background

Most comparison programs are designed to compare non-identical homologous structures, but they can be also used to superpose structures for the same proteins as described in Section 3. There are a large number of different programs and strategies that have been published and we have necessarily had to select just a few as illustrations—our apologies to any author who feels that we have neglected their own work. In general, the methods fall into two different groups:

(a) Those that require the advance definition of pairs of suspected 'equivalent' atoms in order to seed the alignment. An iterative procedure is then used to maximize the number of equivalent matched atom pairs while minimizing the RMSD.

(b) Those methods that sample the realm of possible solutions and, as a result, automatically find optimal alignments without specifying an initial starting alignment.

Some of these procedures involve rigid-body comparisons and others generate a full alignment of the sequences on the basis of the structures. In *Figure 3*, we show the extreme differences in results for the same proteins obtained with a multiple sequence alignment, a rigid-body structure comparison, and a procedure (7) that combines the comparison of properties and relationships to derive the structural matching.

(a) F-STR

```
                *****           ****            **********  ******
4APE-N   ---STGSATTTPIDSLDDAYITPVQ-IGT-----PAQTLNLDFDTGSSDLWVFSSETTASEVDGQTIYTPSK
2APP-N   --AASGVATNTPTA-NDEEYITPVT-IG-------GTTLNLNFDTGSADLWVFSTELPASQQSGHSVYNPSA
2APR-N   --AGVGTVPMTDYG-NDIEYYGQVT-IGT-----PGKKFNLDFDTGSSDLWIASTLCT-NCGSGQTKYDPNQ

4APE-C   YTGSITYTAVSTKQ---GFWEWTSTGYAVGSGTFKSTSIDGIADTGTTLLYLPATVVSA---------YWAQ
2APP-C   YTGSLTYTGVDNSQ---GFWSFNVDSYTAGSQ-SGDG-FSGIADTGTTLLLLDDSVVSQ---------YYSQ
2APR-C   FKGSLTTVPIDNSR---GWWGITVDRATVGTSTVAS-SFDGILDTGTTLLILPNNIAAS---------VARA

                        ****  *
4APE-N   STTAKLLSGATWSISYGDGSSSSGD----VYTDTVSVGGLTVTGQ----------------AVESAKKVS
2APP-N   --TGKELSGYTWSISYGDGSSASGN----VFTDSVTVGGVTAHGQ----------------AVQAAQQIS
2APR-N   SSTYQAD-GRTWSISYGDGSSASGI----LAKDNVNLGGLLIKGQ----------------TIELAKREA

4APE-C   VSGAKSSSSV--------GGYVFPCSA-TLPSFTFGVGSARIVIPGDYIDFGPISTGSSSCFGGIQSSA---
2APP-C   VSGAQQDSNA--------GGYVFDCST-NLPDFSVSISGYTATVPGSLINYGPSGD-GSTCLGGIQSNS---
2APR-C   Y-GASDNGD---------GTYTISCDTSAFKPLVFSINGASFQVSPDSLVFEEF---QGQCIAGFGYG----

                    ****                           ****            ****
4APE-N   SSFTEDSTIDGLLGLAFSTLNTVSPTQQKTFFDNAKAS--LDSPVFTADLGY---HAPGTYNFGFIDTTA
2APP-N   AQFQQDTNNDGLLGLAFSSINTVQPQSQTTFFDTVKSS--LAQPLFAVALKH---QQPGVYDFGFIDSSK
2APR-N   ASFASG-PNDGLLGLGFDTITTVRG--VKTPMDNLISQGLISRPIFGVYLGKAKNGGGGEYIFGGYDSTK

4APE-C   ------GIGINIFGD--------------VALKAA---------FVVFNGA-----TTPTLGFASK----
2APP-C   ------GIGFSIFGD--------------IFLKSQ---------YVVFDSD-----G-PQLGFAPQA---
2APR-C   ------NWGFAIIGD--------------TFLKNN---------YVVFNQG-----V-PEVQIAPVA--E
```

(b) SEQ

```
4APE-N   -STGSATTTPIDSLD-------DAYITPVQIGT-P-AQTLNLDFDTGSSDL----------WVFSSETTAS
2APP-N   AASGVATNTPTAN-D-------EEYITPVTIG----GTTLNLNFDTGSADL----------WVFSTELPAS
2APR-N   AGVGTVPMTDYGN-D-------IEYYGQVTIGT-P-GKKFNLDFDTGSSDL----------WI-ASTLCTN

4APE-C   -YTGSITYTAVSTKQGFWEWTSTGY--AVGSGTFK-STSIDGIADTGTTLLYLPATVVSAYWAQVSGAKSS
2APP-C   -YTGSLTYTGVDNSQGFWSFNVDSYTAGSQSG-----DGFSGIADTGTTLLLLLDDSVVSQYYSQVSGAQQD
2APR-C   -FKGSLTTVPIDNSRGWW----GITVDRATVGTSTVASSFDGILDTGTTLLILPNNIAASV-ARAYGASDN

4APE-N   EVDGQTIYT-PSKSTTAKLLSGATWSISYG-----DGSS---SSGDVYTD--TVSVGGLTVTGQAVESAKK
2APP-N   QQSGHSVYN-P--SATGKELSGYTWSISYG-----DGSS---ASGNVFTD--SVTVGGVTAHGQAVQAAQQ
2APR-N   CGSGQTKYD-PNQSSTYQA DGRTWSISYG-----DGSS---ASGILAKD--NVNLGGLLIKGQTIELAKR

4APE-C   SSVGG--YVFPC-SAT-LP------SFTFG-----VGSARIVIPGD-YIDFGPISTGSSSCFGGIQSSAGI
2APP-C   SNAGG--YVFDC-S-T-N-LPDFSVSIS-GYTATVPGSL--INYGP-SGD------G-STCLGGIQSNSGI
2APR-C   GD-GT--YTI---SCDTSAFKPLVFSI--------NGASFQVSPDSLVFEEFQ---G-QCIAG----F-GY

4APE-N   VSSSFTEDSTIDGLLGLAFSTLNTVSPTQQKTFFDNAKASLDSPVFTADL---GYHAPGTYNFGFIDTTA
2APP-N   ISAQFQQDTNNDGLLGLAFSSINTVQPQSQTTFFDTVKSSLAQPLFAVAL---KHQQPGVYDFGFIDSSK
2APR-N   EAASFASGPN-DGLLGLGFDTITTVRGVKTPMDNLISQGLISRPIFGVYLGKAKNGGGGEYIFGGYDSTK

4APE-C   GINIFG-----DVALKAAF----VVFNGA------------TTP----TL--------G---FASK--
2APP-C   GFSIFG-----DIFLKSQY---VVFD-S------------DGP----QL--------G---FAPQA-
2APR-C   GNWGFAIIG--DTFLKNNY----VVFN-Q------------GVP--------------EVQIAPVAE
```

Figure 3 The differences in alignments of the aspartic proteinase amino- and carboxyl-terminal domains (labelled with an 'N' or 'C', respectively) from (a) multifeature (7) and from (b) multi-sequence comparisons. Asterisks in (a) indicate those positions among the structures that were found to be equivalent under rigid-body superposition with the computer program MNYFIT (16). PDB codes: 4APE, endothiapepsin; 2APP, penicillopepsin; 2APR, rhizopuspepsin. (From ref. 8, with permission.)

4.2 Methods that require the assignment of seed residues

As we have already discussed above, a set of seed matches between a pair of structures is often needed by methods in order to initiate the comparison of structures, because some residue properties, such as $C\alpha$–$C\alpha$ distances, require a partially correct alignment in order to calculate these distances. Once seeded, the alignment improves over several rounds of comparison. Obvious candidates for seed residues are listed in *Protocol 6*.

Protocol 6

Finding initial seed residues

Required data

• Sequences and/or co-ordinates of the proteins to be compared

Method

1 Supply a minimum of three conserved residues from a sequence-based alignment, or

2 Supply key residues implicated in a conserved binding or catalytic motif, or

3 Supply segments corresponding to secondary structure elements observed on a graphics device to be conserved between the structures.

In *Protocol 7*, we present a general procedure for the alignment of two structures using rigid-body comparisons, which requires a seed set of matches between the two 3-D structures.

Protocol 7

Semi-automatic methods

Required data

• Co-ordinates of the proteins to be compared

• Initial set of equivalent atom pairs to seed the alignment procedure

Method

1 Calculate translation vector based on seed residues, translate all co-ordinates to the origin and calculate the rotation matrix for the seed residues (see Protocol 2).

2 Apply the rotation matrix to all atoms of the second protein to achieve the first superposition (see Protocol 2).

3 Obtain the alignment using dynamic programming or clique analysis (see Protocol 3).

4   For all matched residue pairs in the alignment, calculate the Euclidean distance. Those matched pairs within the distance cut-off value will form the new updated set of equivalent atom pairs for the next round of super-positioning.

5   Repeat steps 1–4 until convergence is obtained: the calculated RMSD (Protocol 4) does not decrease and the number of equivalent atom pairs matched in the two proteins does not increase.

6   A rigid-body comparison has been used as a starting point for more detailed structural comparisons involving multiple structural features (e.g. the program COMPARER described in ref. 7).

4.3 Automatic comparison of 3-D structures

To get around the requirement for an initial set of equivalent seed matches, alternative methods have been developed. Here, several of the many published methods are described to illustrate the different strategies that have been employed:

(a) Methods that supply seed matches automatically to a rigid-body approach after making a sequence-based alignment (*Protocol 8*).

(b) Methods that use a genetic algorithm (*Protocol 9*) or 'spectra'-comparison method (*Protocol 10*) to find the optimal rigid-body comparison.

Protocol 8

Structural comparisons seeded from sequence alignments

Automatic alignment of two homologous protein structures without the need to specify initial equivalent atoms pairs. Method can fail for proteins of low sequence similarity.

Required data

• Cα-atom co-ordinates of the compared structures and their sequences

Method

1   Align the amino acid sequences with a dynamic programming algorithm (Protocol 3, but using sequence-matching scores to produce the alignment).

2   Superimpose the structures according to Protocol 7 using the most conserved portions of the sequence alignment as the initial set of seed residues.

(c) Methods that do not make rigid-body comparisons directly, but instead make comparisons on the basis of similarities in structural properties and/or relationships (*Protocols 11–14*).

4.3.1 Structural comparisons seeded from sequence alignments

Russell and Barton (9) developed a method that first makes an alignment of the sequences and then uses equivalent matches defined in the alignment to seed a structural comparison. We have given a general protocol for such an approach above (*Protocol 8*). The procedure should work well for proteins where portions of the sequence alignment can be trusted; when the sequence similarity is low and the alignment is not correct, then the method may not be useful.

4.3.2 Rigid-body comparisons using a genetic algorithm

Genetic algorithms (29) describe the solution to a problem within a numerical string. A large number of strings are originally assigned random values as their solutions, and the genetic algorithm seeks to evolve this initial set towards better and better solutions by exchanging partially good solutions among strings and by mutating the strings. *Protocol 9* describes the general procedures used by May and Johnson (26, 27) to automatically compare one or more structures. The approach is time-consuming but has been successfully adapted to parallel processors (Lehtonen and Johnson, unpublished).

Protocol 9
GA_FIT (ref. 26, 27)

Automatic rigid-body alignment of two protein structures without the need to specify initial equivalent atoms pairs.

Required data
• Cα-atom co-ordinates of compared structures

Method

1 Create a large random set of superpositions for the pair of structures.

2 Assign equivalent matches (Cα-atoms within a specified distance cut-off) using dynamic programming and score each alignment (see Protocol 3).

3 Create a new set of superpositions by crossing-over and mutating the existing solutions.

4 Repeat steps 2–3 until a close to final solution is achieved.

5 Optimize the best found superposition/alignment by least squares minimization (see Protocol 2).

6 Calculate the final alignment with the dynamic programming algorithm (see Protocol 3).

4.3.3 Rigid-body comparisons using the density of Cα-atom packing and spectral alignment

VERTAA, described in *Protocol 10* (Lehtonen and Johnson, unpublished method), compares and aligns spectra equal to the Cα-atom density in each structure as a function of the position along the sequence of each protein (*Figure 4*). This method is a rapid and automatic means for comparing structures.

Protocol 10

Local similarity search by VERTAA

Fast, automatic alignment of two protein structures without the need to specify initial equivalent atoms pairs.

Required data

• Cα-atom co-ordinates of the two structures.

Method

1   VERTAA, for each of two structures, plots the number of Cα-atoms within a given radius (14.0 Å) from each Cα-atom in the structure. Other properties can be used too.

2   These 'spectra' are scaled and overlapping segments are aligned. More than one alignment method is available:

   (a) The dynamic programming algorithm (Protocol 3). Fast and robust if the input values are properly scaled.

   (b) The Fourier correlation (21). The values can be considered as a function over a limited range and a correlation function obtained with the fast Fourier transform to bring the spectra into register. Dynamic programming is then used to define equivalent matches (Protocol 3).

3   Superimpose the structures (see Protocol 2) based on equivalent matches defined in step 2.

4   Define a new alignment with dynamic programming and the Cα-Cα distances of the superimposed structures within 3.5 Å (see Protocol 3).

5   While the alignment and superimposition improve, repeat steps 3 and 4.

4.3.4 Structural comparisons based on matching Cα-atom contact maps

Holm and Sander (11) make comparisons by comparing Cα atom–Cα atom contact maps (by contacts, we mean nearby in space) constructed from each protein structure (*Protocol 11*).

(a)

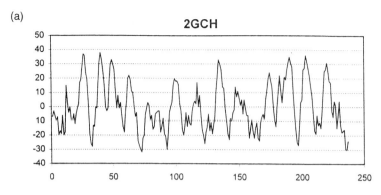

(b)

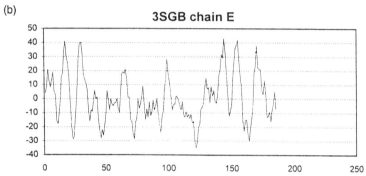

(c)

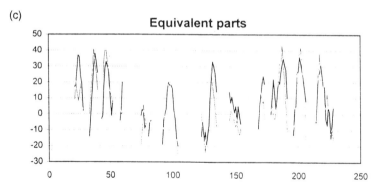

(d)

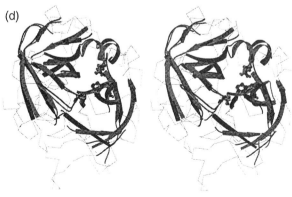

Figure 4 Plots of Cα-atom densities, alignment of plots, and the corresponding superposition of the structures. (a) Cα-atom densities of residues in γ chymotrypsin A (PDB code 2GCH). (b) Cα-atom densities of residues in *Streptomyces griseus* proteinase B (PDB code 3SGB, chain E). For both spectra, the average density is set equal to 0. (c) The parts of the two plots from (a) (dark) and (b) (light), which correspond to each other according to the alignment of their spectra. (d) The superpositioned 3-D structures (2GCH dark, 3SGB light) based on the alignment specified in (c). The side chains of the catalytic triad are shown and the closely matching parts are drawn as ribbon diagrams. This superposition was made with the computer program VERTAA (Lehtonen and Johnson, unpublished results) and contains 118 residues within 3.5 Å with an RMSD of 1.8 Å. Figure (d) was prepared with MOLSCRIPT (52).

Protocol 11

Structure comparison by DALI (11)

Automatic alignment by finding the optimal clique for contact maps obtained from the structures (Protocol 3).

Required data

- Cα-atom co-ordinates of compared structures

Method

1 Calculate a distance matrix for each protein A. Element (i, j) of the matrix contains the intramolecular distance between the ith and j^{th} Cα-atom in A. Likewise, calculate a distance matrix for protein B.

2 List from each distance matrix all possible 6 by 6 sub-matrices.

3 Reduce the number of sub-matrices by clustering similar ones and using the mean of each cluster as the contact pattern. Sort contact patterns by intra-pattern distance.

4 Compare each pair of two contact patterns from A with all pairs of sub-matrices from B. Compare each pair of two contact patterns from B with all pairs of sub-matrices from A. List all pair–pair matches.

5 Remove redundancy from the list of matches and sort it by match quality, which is a function of the differences between the sub-matrices from A and from B.

6 Find the most extensive, non-exclusive collection of matches from the list. DALI uses a Monte Carlo simulation to search the best 40 000 matches. The simulation tries to extend the matches by combining matches that contain a common contact pattern in both distance matrices. The random element of the simulation is used to find the best scoring combination from mutually exclusive possibilities.

4.3.5 Comparisons using double dynamic programming

Taylor and Orengo (10) have developed a novel use of dynamic programming in order to facilitate the comparison of relationships. Dynamic programming is used once to compare structural relationships in the two proteins thus providing scores for a second round of dynamic programming where the two structures are aligned (*Protocol 12*).

Protocol 12

Structure comparison by SSAP (10)

Fast automatic alignment of two protein structures using double dynamic programming.

Required data

• Cα-atom co-ordinates of compared structures

Method

1   Calculate a distance matrix for protein A. Element (i, j) of the matrix contains the intramolecular vector from the ith to the jth Cα-atom in A. The vector is in the co-ordinate frame defined by the covalent bonds of A's i^{th} Cα-atom. Likewise, calculate a distance matrix for protein B.

2   Calculate intramolecular difference matrices for each pair of rows from the two distance matrices. Thus, element (i, j) of the matrix constructed from row h of A's, and row k of B's distance matrix will contain the difference of the magnitude of the vectors $\overset{\omega}{A}_{hi}$ and $\overset{\omega}{B}_{ki}$ converted to a similarity value.

3   Low level alignments of the local structure are made first using a dynamic programming algorithm (see Protocol 3) to find the best scoring path through the intramolecular difference matrices (see ref. 10 for details). Scores along the path will contribute to a separate 'summed scoring matrix' from which the final alignment will be determined.

4   Use a dynamic programming algorithm (see Protocol 3) to trace an alignment path through the summed scoring matrix. This higher level alignment defines the equivalent matches between the structures.

4.3.6 Structural alignments based on secondary structure element (SSE) matching

Kleywegt and Jones (30) describe a method for structural comparisons based on the alignment of elements of regular secondary structure (*Protocol 13*).

Protocol 13

Structure comparison by DEJAVU (30)

Automatic alignment of protein structures by finding the optimal clique on the basis of secondary structure comparisons (Protocol 3).

Required data

• Cα-atom co-ordinates of the two structures or SSE templates of the structures

Generation of SSEs with YASSPA in O (see ref. 30 for details)

Search the structures and tabulate main chain fragments that are similar to templates of typical α-helices and β-strands.

Protocol 13 continued

Comparison of structures with DEJAVU (see ref. 30 for details)

1 Check that both structures have the required number of SSEs.

2 Check that there exists at least one SSE of the same type (same length—number of Cα-atoms) in the second structure for each SSE in first structure.

3 Find the most extensive, non-exclusive collections of matched SSEs. DEJAVU does a depth-first tree search to find all sets of matching SSE pairs, where all pairs in a set are matching also in 3-D space. The tree contains all possible combinations of pairs. If the path from the root to a node already has too many mismatches, the sub-tree below the node is not searched, saving time.

4 Report the matched SSEs and the Cα-atoms for the best scoring alignment.

5 The output can be directed to external programs for refinement of the super-position and visualization.

4.4 Multiple structural comparisons

Multiple structural comparisons can be made using several different strategies. Sutcliffe *et al.* (16) constructed multiple rigid-body structural alignments by comparing each structure to an average representation of the structures (in practice, one of the structures was chosen for this purpose at the beginning of the comparisons). More frequently, multiple alignments are assembled from pairwise structural alignments according to the topology of a tree estimated on the basis of sequence alignments (9, 27, *inter alia*). This (*Protocol 14*) follows the strategy first introduced by Barton and Sternberg (31) and Feng and Doolittle (32) for the efficient multiple alignment of protein sequences.

Protocol 14

Multiple structural alignments from pairwise comparisons

Multiple alignments assembled from pairwise comparisons.

Required data

• Cα-atom co-ordinates of the structures in PDB format

A general approach

1 Use a sequence alignment procedure to align the proteins and to cluster them as a bifurcating tree (see refs 31–34 and several chapters in ref. 35).

2 Use a pairwise structural alignment method to align clusters according to the tree topology. This will involve comparing pairs of structures, one structure with a set of previously aligned structures, and aligned structures with aligned structures, until all clusters have been coalesced into a final alignment involving all of the proteins.

5 The comparison of unrelated structures

5.1 Background

Non-homologous protein structures have frequently been compared to high-light features of protein structure that are common across many families. It has been less often recognized, however, that proteins with different folds can also share similarities that can extend to a fairly large organization of their structures, for example about common ligand and cofactor binding sites. The elements contributing to these similarities are likely to involve fragments of each structure that do not map along the protein chains in any predetermined way (*Figure 5* and ref. 22). Thus, despite similar local structure, the segments contributing to the similarity can be both rearranged and discontinuous with respect to each other (e.g. 36–38). Such similarities are particularly difficult to recognize even if a hint of a common functional requirement is present, like a common cofactor. Nonetheless, the recognition of local similarities can provide evidence about the rules governing the structure-function relationship suitable for protein modelling, the prediction of structure from sequence and computer-based drug design. For example, Kobayashi and Go (39) have reported a local motif about the ATP binding site common to cyclic-AMP dependent protein kinase and D-Ala:D-Ala ligase involving 4 equivalent residues. Comparisons using the computer program GENFIT (22) automatically and repeatedly found up to 60 matches (36) that includes an extensive supersecondary structure organization used to position polar and nonpolar residues that interact with the similarly oriented cofactor, bound metal and bound water molecules (*Figure 6*).

Given two unrelated protein structures, A and B, the goal of a computer program is to find the largest equivalent subset of the two structures. Because the proteins are not derived from a common ancestor, the matches providing equivalent structural interactions:

(a) Are not necessarily sequential along the two sequences (*Figure 5*).

(b) Can involve matched elements of secondary structure whose chain directions are opposite to each other (*Figure 5*) but can still provide equivalent interactions, for example, with bound ligand.

Here we describe two different approaches that have been successfully used to find similarities among unrelated protein structures, SARF2 (40, 41) and GENFIT (22). SARF2 (41) considers SSEs, finds the maximal common sub-graph for two structures; and systematically creates different alignments, tries to improve them, evaluates them and reports the best found alignments (*Protocol 15*). GENFIT (22) considers matched segments of Cα-atoms and employs a genetic algorithm to randomly sample large numbers of possible alignments and uses a match-list approach to assign equivalent segments of structure, which are subsequently used to make a local rigid-body superposition for each alignment (*Protocol 16*). GENFIT, by virtue of the genetic algorithm, will find and report different equally likely superpositions in different runs (*Figure 7*).

Both of these approaches establish equivalent matches between objects,

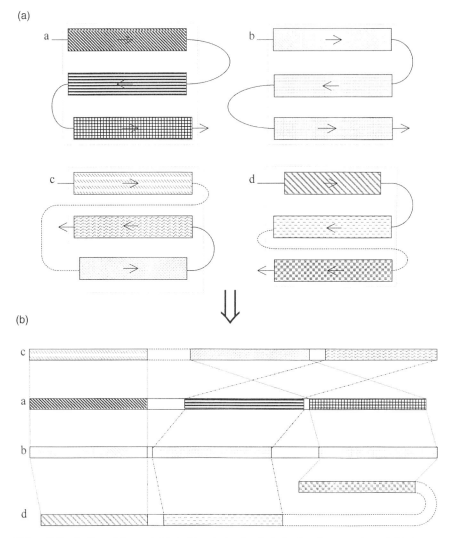

Figure 5 Illustration of the matching of local structural similarities from non-homologous proteins *versus* homologous proteins. (a) Topological diagrams show four protein structures (*a*, *b*, *c*, and *d*) with similar local structural elements. Topologies *a* and *b* represent two homologous structures with the same fold, while *c* has a different topology than *a* and *b*, yet has the same core structure. Topology *d* illustrates a different fold that still has the structurally equivalent segments of polypeptide chain in same place, but some segments may have opposite chain directions. In (b), the correspondence found in the structural alignment is shown at the sequence level. Note that only the sequences of *a* and *b* have a straightforward linear correspondence. (From ref. 22, with permission.)

segments of Cα-atoms (GENFIT) or SSEs (SARF2). GENFIT starts with 'too many' equivalent matches and reduces them until a maximal, but non-conflicting set, is obtained. This is done for each of the many parallel comparisons being made, but a single optimal result will be obtained in any one run: the parallel com-

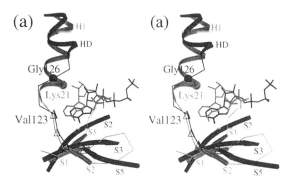

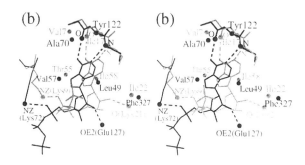

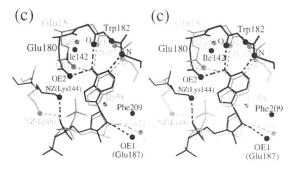

Figure 6 Local similarity of ATP cofactor binding site seen in the pairwise superposition of ribonucleotide reductase (PDB code: 3R1R, chain A; light grey) with cAMP-dependent protein kinase (1CDK; dark grey) (a and b), and with D-Ala:D-Ala ligase (1IOW; dark grey) (c). The bound ATP molecules of the structures are shown as stick models. In (a), the four common segments are drawn as ribbon diagrams. (b and c) The environment around the cofactor is illustrated by showing the equivalent hydrogen bonds (dashed lines) and equivalent C$^\alpha$-atoms (spheres) forming hydrophobic contacts to the cofactor. (From ref. 37, with permission.)

parisons converge towards that result. SARF2 searches among a large set of matches between the structures and finds the largest non-conflicting subset of matches. Both methods are free from restraints on the order and chain direction of objects along the sequence, but optional restraints can be applied.

Protocol 15

Structure comparison by SARF2 (41)

Local similarity alignment of non-homologous structures.

Required data

- Cα-atom co-ordinates of the two structures

Method

1   Search for and tabulate main-chain fragments from the structures that are similar to five-residue long templates of typical α-helices and β-strands.

2   Create a list of SSE pairs from the first structure that match SSE pairs from the second structure. Distance and angular criteria between the SSE's in both structures is important to the determination of a match.

3   Combine matches to find the largest collection of SSE's that can be aligned. SARF2 uses an exhaustive, recursive search algorithm to find possible solutions (see ref. 41 for details).

4   For the best solutions found, superimpose the matched SSE's and then add nearby Cα-atoms to matched regions using the dynamic programming method. Iteratively repeat the superpositions of Cα-atoms until the maximum number of matched atoms have been found.

5   A list of superpositions, ranked according to an alignment score, result.

Protocol 16

GENFIT (22)

Automatic alignment of two locally similar protein structures using a genetic algorithm. This implementation has been designed for parallel processing environments.

Required data

- Cα-atom co-ordinates of the two structures

Method

1   Create a large random set of superpositions for the pair of structures.

2   Assign equivalent matches using the match list algorithm (see Protocol 3). Criteria for a match include:

   (a) Cα-atom matches must be within a user specified distance cut-off.

   (b) Matches must include a minimum of four consecutive Cα-atoms.

   (c) The direction of the main chain for matched segments is unimportant by default.

   (d) Matches do not need to be co-linear (i.e. the location of a match along the sequence relative to other matches is unimportant).

Protocol 16 continued

3   Calculate an alignment score for each superposition and create a new set of superpositions by crossing-over and mutating existing ones (see ref. 22 for details).

4   Repeat steps 2 and 3 until convergence has been achieved.

5   Optimize the best superposition/alignment by least-squares rigid-body minimization (Protocol 2).

6   Recalculate the alignment with the match list algorithm (Protocol 4).

7   If the number of equivalent matches has increased or the fit has improved, repeat steps 5 and 6 with the current alignment.

8   Repetitive runs can produce different results showing that equally likely alternative results exist.

6 Large-scale comparisons of protein structures

One of the straightforward goals in bioinformatics today is to compare, cluster and classify both sequences and the known 3-D structures. Initially, this means categorizing each existing sequence or structure in a data bank. Then, when new entries are made to sequence and structure data banks, each new entry will need to be compared against the existing classifications.

The methods described in this chapter can and have been applied to such analyses. For example, both MNYFIT (16) and COMPARER (7) have been used to accurately align all families of 3-D structures containing two or more structures (43–45), that can be accessed in a public database: *http://www–cryst.bioc.cam.ac.uk/ cgi-bin/joy.cgi*. Other available databases include FSSP (46) created using DALI (11): *http://www.ebi.ac.uk/dali/fssp/*; and CATH (47) created in part using SSAP (10): *http://www.biochem.ucl.ac.uk/bsm/cath/*. Several other data banks worth mentioning include MMDB (48): *http://www.ncbi.nlm.nih.gov/Structure/* and SCOP (49): *http: //scop.mrc-lmb.cam.ac.uk/scop/*

In *Figure 8*, we present a classification of structures from several different families that belong to the all-β structural classification. This classification was made by comparing the structures on the basis of their secondary structures and then clustering them according to the pairwise structural similarity (44, 50).

Figure 7 Two examples of differing alignments of locally similar structures. (a and b) Superposition of UDP-galactose 4-epimerase chain B (2UDP) and DNA methyltransferase (1HMY) showing similarity between the larger domains. In (b), 1HMY has been rotated by 180 degrees around the axis of the β-sheet in comparison to (a). The symmetry of the nearly planar β-sheet allows for several different, but similarly-scoring alignments. (c and d) Superposition of cyclic-AMP-dependent protein kinase (1CDK) and glutamine synthetase (1LGR) showing local similarities about the ATP-binding sites and the differences seen from matching fewer longer segments (α-helices) or many shorter segments (β-sheets). In (c), the antiparallel β-sheets are aligned and the cofactors overlap, while in (d) the α-helices are matched, but the β-sheets and the cofactor do not superpose well. The superpositions have been made with program GENFIT (22). (From ref. 22, with permission.)

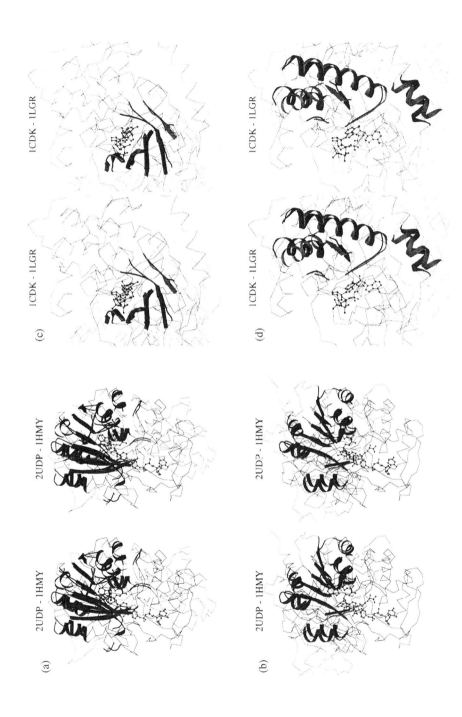

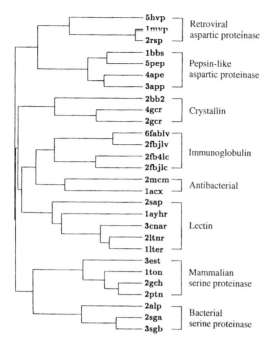

Figure 8 Dendogram of clusters of protein structures composed primarily from β-strands. Each cluster, a family of proteins, is distinguished from the others by its unique fold. (From ref. 44, with permission.)

References

1. Dayhoff, O. M., Barker, W. C., and Hunt, L. T. (1983). In *Methods in enzymology* (ed. C. H. W. Hirs and S. W. Timasheff). Vol. 91, p. 524. Academic Press, London.
2. Chothia, C. (1992). *Nature*, **357**, 543.
3. Blundell, T. L. and Johnson, M. S. (1993). *Protein Sci.*, **2**, 877.
4. Johnson, M. S., Srinivasan, N., Sowdhamini, R., and Blundell, T. L. (1994). *Crit. Rev. Biochem. Mol. Biol.*, **29**, 1.
5. Robertus, J. D., Alden, R. A., Birktoft, J. J., Kraut, J., Powers, J. C., and Wilcox, P. E. (1972). *Biochemistry*, **11**, 2449.
6. Needleman, S. B. and Wunsch, C. D. (1970). *J. Mol. Biol.*, **48**, 443.
7. Sali, A. and Blundell, T. L. (1990). *J. Mol. Biol.*, **212**, 403.
8. Johnson, M. S., Sali, A., and Blundell, T. L. (1990). In *Methods in enzymology* (ed. R. F. Doolittle), Vol. 183, p. 670. Academic Press, San Diego.
9. Russell, R. B. and Barton, G. J. (1992). *Proteins*, **14**, 309.
10. Taylor, W. R. and Orengo, C. A. (1989). *J. Mol. Biol.*, **208**, 1.
11. Holm, L. and Sander, C. (1993). *J. Mol. Biol.*, **233**, 123.
12. McLachlan, A. D. (1972). *Acta Crystallogr.*, **A28**, 656.
13. McLachlan, A. D. (1979). *J. Mol. Biol.*, **128**, 49.
14. McLachlan, A. D. (1982). *Acta Crystallogr.*, **A38**, 871.
15. Diamond, R. (1988). *Acta Crystallogr.*, **A44**, 211.
16. Sutcliffe, M. J., Haneef, I., Carney, D., and Blundell, T. L. (1987). *Protein Eng.*, **1**, 377.
17. Diamond, R. (1992). *Protein Sci.*, **1**, 1279.

18. Shapiro, A., Botha, J. D., Pastor, A., and Lesk, A. M. (1992). *Acta Crystallogr.*, **A48**, 11.
19. Kearsley, S. K. (1989). *Acta Crystallogr.*, **A45**, 208.
20. Kearsley, S. K. (1990). *J. Comput. Chem.*, **11**, 1187.
21. Press, W. H., Teukolsky, S. A., Vetterling, W. T., and Flannery, B. P. (ed.) (1992). *Numerical recipes in C. The art of scientific computing.* (2nd edn). Cambridge University Press, Cambridge.
22. Lehtonen, J. V., Denessiouk, K., May, A. C. W., and Johnson, M. S. (1999). *Proteins*, **34**, 341.
23. Mitchell, E. M., Artymiuk, P. J., Rice, D. W., and Willett, P. (1990). *J. Mol. Biol.*, **212**, 151.
24. Gibrat, J.-F., Madej, T., and Bryant, S. H. (1996). *Curr. Opin. Struct. Biol.*, **6**, 377.
25. Grindley, H. M., Artymiuk, P. J., Rice, D. W., and Willet, P. (1993). *J. Mol. Biol.*, **229 (3)**, 707.
26. May, A. C. W. and Johnson, M. S. (1994). *Protein Eng.*, **7**, 475.
27. May, A. C. W. and Johnson, M. S. (1995). *Protein Eng.*, **8**, 873.
28. Johnson, M. S., Sutcliffe, M. J., and Blundell, T. L. (1990). *J. Mol. Evol.*, **30**, 43.
29. Goldberg, D. E. (ed.) (1989). *Genetic algorithms in search, optimization, and machine learning.* Addison-Wesley, Reading, MA.
30. Kleywegt, G. J. and Jones, T. A. (1997). In *Methods in enzymology* (ed. C. W. Carter and R. M. Sweet), Vol. 277, p. 525. Academic Press.
31. Barton, G. J. and Sternberg, M. J. (1987). *J. Mol. Biol.*, **198**, 327.
32. Feng, D. F. and Doolittle, R. F. (1987). *J. Mol. Evol.*, **25**, 351.
33. Johnson, M. S. and Overington, J. P. (1993). *J. Mol. Biol.*, **233**, 716.
34. Johnson, M. S., May, A. C. W., Rodionov, M. A., and Overington, J. P. (1996). In *Methods in enzymology* (ed. R. F. Doolittle, Vol. 266, p. 575. Academic Press.
35. Doolittle, R. F. (ed.) (1996). In *Methods in enzymology*, Vol. 266, p. 711. Academic Press, San Diego.
36. Denessiouk, K., Lehtonen, J. V., Korpela, T., and Johnson, M. S. (1998). *Protein Sci.*, **7**, 1136.
37. Denessiouk, K., Lehtonen, J. V., and Johnson, M. S. (1998). *Protein Sci.*, **7**, 1768.
38. Denessiouk, K., Denesyuk, A. I., Lehtonen, J. V., Korpela, T., and Johnson, M. S. (1999). *Proteins*, **35**, 250.
39. Kobayashi, N. and Go, N. (1997). *Nature Struct. Biol.*, **4**, 6.
40. Alexandrov, N. N., Takahashi, T., and Go, T. (1992). *J. Mol. Biol.*, **225**, 5.
41. Alexandrov, N. N. and Fischer, D. (1996). *Proteins*, **25**, 354.
42. Sali, A., Overington, J. P., Johnson, M. S., and Blundell, T. L. (1990). *Trends Biochem. Sci.*, **15**, 235.
43. Overington, J. P., Johnson, M. S., Sali, A., and Blundell, T. L. (1990). *Proc. R. Soc. Lond. B*, **241**, 132.
44. May, A. C. W., Johnson, M. S., Rufino, S. D., Wako, H., Zhu, Z.-Y., Sowdhamini, R., *et al.* (1994). *Phil. Trans. R. Soc. Lond. B*, **344**, 373.
45. Mizuguchi, K., Deane, C. M., Blundell, T. L., Johnson, M. S., and Overington, J. P. (1998). *Bioinformatics*, **14**, 617.
46. Holm, L. and Sander, C. (1998). *Nucleic Acids Res.*, **26**, 316.
47. Orengo, C. A., Michie, A. D., Jones, S., Jones, D. T., Swindells, M. B., and Thornton, J. M. (1997). *Structure*, **5**, 1093.
48. Marchler-Bauer, A., Addess, K. J., Chappey, C., Geer, L., Madej, T., Matsuo, Y., *et al.* (1999). *Nucleic Acids Res.*, **27**, 240.
49. Murzin, A. G., Brenner, S. E., Hubbard, T., and Chothia, C. (1995). *J. Mol. Biol.*, **247**, 536.
50. Rufino, S. D. and Blundell, T. L. (1994). *J. Comput. Aided Mol. Des.*, **8**, 5.

51. Bernstein, F. C., Koetzle, T. F., Williams, G. J. B., Meyer, E. J. Jr, Brice, M. D., Rodgers, J. K., *et al.* (1977). *J. Mol. Biol.*, **112**, 535.
52. Kraulis, P. J. (1991). *J. Appl. Crystallogr.*, **24**, 946.

Chapter 3

Multiple alignments for structural, functional, or phylogenetic analyses of homologous sequences

L. Duret

Laboratorie de Biométrie, Génétique et Biologie des Populations, Université Claude Bernard, France.

S. Abdeddaim

Département d'Informatique de Rouen, Universite de Rouen, France.

1 Introduction

Understanding the structure, function and evolution of genes is one of the main goals of genome sequencing projects. Classically, gene function has been investigated experimentally through the analysis of mutant phenotypes. More recently, comparative analysis of homologous sequences has proved to be a very efficient approach to study gene function (this approach has been coined 'comparative genomics' or 'phylogenomics'). Indeed, the evolution of living organisms may be considered as an ongoing large-scale mutagenesis experiment. For more than three billion years, genomes have continuously undergone mutations (substitutions, insertion, deletions, recombination, and so on). Deleterious mutations are generally rapidly eliminated by natural selection, while mutations that have no phenotypic effect (neutral mutations) may, by random genetic drift, eventually become fixed in the population. Globally, advantageous mutations are very rare, and hence residues that are poorly conserved during evolution generally correspond to regions that are weakly constrained by selection (1). Thus, studying mutation patterns through the analysis of homologous sequences is useful not only to study evolutionary relationships between sequences, but also to identify structural or functional constraints on sequences (DNA, RNA, or protein).

The alignment of homologous sequences consists of trying to place residues (nucleotides or amino acids) in columns that derive from a common ancestral residue. This is achieved by introducing gaps (which represent insertions or deletions) into sequences. Thus, an alignment is a hypothetical model of mutations (substitutions, insertions, and deletions) that occurred during sequence evolution. The best alignment will be the one that represents the most likely evolutionary scenario. Generally, this best alignment cannot be unambiguously established. Firstly, because of the computational complexity of this problem, alignment algorithms that are usable in practice cannot guarantee to find the best solution (see *Section 4.1*). Secondly, even with an ideal algorithm, finding the best alignment would not be guaranteed because current knowledge of the probability of occurrence of the different types of mutation is still limited (see *Section 2.3*). However, as long as homologous sequences are not too divergent, fast approximate algorithms may be used to provide reliable alignments. In practice, such alignments are commonly used in molecular or evolutionary biology. Typical examples of usage of multiple alignments are indicated in *Table 1*.

Table 1 Examples of usage of multiple alignments

• **Identification of functionally important sites**
Multiple alignments allow the identification of highly conserved residues are likely to correspond to essential sites for the structure or function of the sequence and may thus be useful to design mutagenesis experiments.

• **Demonstration of homology between sequences (see Section 2.1)**

• **Molecular phylogeny**
Molecular phylogenetic trees rely on multiple alignments (protein or DNA) to infer mutation events from which it is possible to retrace evolutionary relationships between sequences. Such trees are useful to reconstruct the history of species or multigenic families, and notably to identify gene duplication events to distinguish orthologues from paralogues. It is important to note that unreliable parts of alignments should not be used to build phylogenetic trees since they do not reflect the real pattern of mutations that occurred during evolution and may lead to artifactual results.

• **Search for weak but significant similarities in sequence databases**
The sensitivity of sequence similarity search may be improved by weighting sites according to their degree of conservation. Thus, multiple alignments of homologous sequences are used by methods such as profile searches (see the chapter by Henikoff in this volume) or PSI-BLAST (24) to identify distantly related members of a family.

• **Structure prediction**
The use of multiple alignments increases significantly the efficiency of protein secondary structure prediction. Moreover, the identification of covariant sites (or compensatory mutations) in alignments (protein or RNA) is a strong argument to suggest that these sites interact in the molecule *in vivo*. Finally, alignments are commonly used for homology modeling, i.e. for the structure prediction of sequences by comparison with homologues of known structure.

• **Function prediction**
The three-dimensional (3D) structure of homologous proteins or RNA is often much more conserved than their primary sequence. Similar shape usually implies similar function. Thus, if a new gene is found to be homologous to an already characterized gene it is possible to infer the likely function of the new gene from the known one. Such inferences should however be used with great caution.

• **Design of primers for PCR (polymerase chain reaction) identification of related genes**

The general procedure to compute a multiple alignment of homologous sequences consists of three steps:

(a) Search for homologues in sequence databases.

(b) Compute alignments.

(c) Check and edit alignments.

In this chapter, we will focus, essentially, on steps (b) and (c). Firstly, we will define some general concepts underlying multiple alignment methodology. We will then describe and compare different methods that have been developed to align sequences. As far as possible, we will indicate WWW sites where these tools are available, so that they may be used from any computer with an appropriate WWW browser software and internet connection. The list of WWW sites that we provide here is also available at the following address:

http://pbil.univ-lyon1.fr/alignment.html

Some problems such as contig assembly, related to multiple alignment will not be treated in this chapter. We will only describe methods intended for the alignment of homologous sequences and, in particular, we will not deal with the problem of finding common motifs in a set of unrelated sequences. Note that there is not an absolute difference between motif search and multiple alignments: when homologous sequences have diverged too much there may remain only a few short conserved fragments, separated by regions of variable length. Motif-based methods have been developed to identify and align such conserved fragments within highly divergent sequences. In *Section 4.4* we will mention some of these methods. However, for a more exhaustive review on this topic, see Chapter 7 by Jonassen in this volume.

2 Basic concepts for multiple sequence alignment

2.1 Homology: definition and demonstration

Two sequences are said to be homologous if they derive from a common ancestor. Generally, homology is inferred by sequence similarity. It should be stressed, however, that similarity does not necessarily reflect homology: similarity between short sequence fragments may result from evolutionary convergence (2), or may simply occur by chance. Moreover many sequences contain relatively long fragments of very biased nucleotide or amino acid composition (e.g. CA-repeats in DNA, proline-rich domains in proteins) (3). Generally, similarities between such 'low complexity regions' do not reflect evolutionary relationship. However, in the absence of such compositional bias, similarity over an extended region usually implies homology. Statistical tests can be used to evaluate the chances that an observed similarity occurred purely by chance and thus accept or reject the hypothesis of homology (4). Such tests are now generally provided by similarity search programs.

Multiple alignments may be useful to help demonstrate homology: a weak similarity which would be considered as non-significant in a pairwise sequence comparison may prove to be highly significant if the same residues are conserved in other distantly related sequences. It should be emphasized that if sequences have diverged too much, homology may not be recognizable on the basis of sequence similarity alone.

2.2 Global or local alignments

In the above paragraph, we implicitly considered sequences that are homologous over their entire length. However, in many cases, homology is restricted to a limited region of the sequences. Indeed, many proteins consist of a combination of discrete 'modules' that have been shuffled during evolution. It is clear that many protein-coding genes result from recombination between different fragments of other genes. This modular evolution has played a major role in protein evolution and has been particularly facilitated in eukaryotes thanks to the presence of introns within genes (5).

Multiple copies of a given module may be repeated within a sequence, and a set of modules may occur at different relative positions in different genes. In such cases, it is not possible to align sequences over their whole length (global alignment) and it is thus necessary to perform alignments only on homologous modules (local alignment) See *Figure 1* for an illustration.

2.3 Substitution matrices, weighting of gaps

As indicated earlier, searching for the best alignment consists of searching for the one that represents the most likely evolutionary scenario. Thus, the prob-

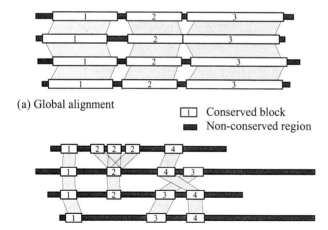

(a) Global alignment

☐ Conserved block
▬ Non-conserved region

(b) Local alignment

Figure 1 Global versus local alignment. (a) Conserved regions occur in the same order in all sequences. They can be represented in a single global alignment. (b) Some conserved regions are duplicated or occur in a different order along sequences. It is necessary to perform local alignments to display similarities between all conserved regions.

ability of occurrence of the different mutational events during evolution must be taken into consideration when computing a multiple alignment. In alignments, three types of mutations are considered: substitutions, insertions or deletions (the two latter events are often indistinguishable, and are commonly referred as 'indels').

2.3.1 Substitutions

The probability of substitution of one amino acid by another depends on the structure of the genetic code (i.e. on the number of mutations necessary to pass from one codon to another) and also on the phenotypic effect of that mutation. Substitutions of one amino acid by another with similar biochemical properties generally do not greatly affect the structure and hence the function of the protein. Thus, during evolution, such conservative substitutions are relatively frequent compared to other substitutions. It is important to note that the probability of substitution of one amino acid by another depends on the evolutionary distance between sequences. At short evolutionary distances, probabilities of substitution mainly reflect the structure of the genetic code, whereas at larger distances, probabilities of substitution depend essentially on biochemical similarities between amino acids. Various methods have been proposed to build series of matrices that give estimates of probabilities of all possible substitutions for different evolutionary distances (6–8). The most commonly used are the PAM and BLOSUM substitution matrices. PAM matrices suitable for increasing evolutionary distances are indicated by increasing indices (e.g. PAM80, PAM120, and PAM250). The opposite convention has been used for the BLOSUM series (e.g. BLOSUM80 for short evolutionary distances, BLOSUM45 for large evolutionary distances). Generally, alignment programs allow users to choose which substitution matrix to use. In the CLUSTAL W program (9) (see *Section 4.2*) substitution matrices are automatically selected and varied at different alignment stages according to the divergence of the sequences to be aligned.

Probabilities of substitutions also vary along sequences according to the local environment of amino acids in the folded protein. Thus, several environment-specific substitution matrices have been developed (e.g. for α-helix, or β-sheet) (10). However, to our knowledge, these matrices are rarely used for multiple alignments.

At the DNA level, probabilities of substitution vary according to the bases. Notably, transitions (substitutions between two purines—A, G—or two pyrimidines—C, T) are generally more frequent than transversions (substitutions between a purine and a pyrimidine). Thus, multiple alignment programs generally propose a parameter to weight more heavily transversions than transitions. Probabilities of nucleotide substitution also depend on neighbouring bases (e.g. in vertebrates, C in CG dinucleotides is hypermutable) (11, 12). However, currently available alignment programs do not make use of such information.

2.3.2 Insertions, deletions

The probability of occurrence of an indel depends on its length. Thus, when computing an alignment, penalties (p) associated with gaps are often estimated using a linear or 'affine' model such as:

$$p = a + bL$$

where L is the length of the gap, a the gap opening penalty, and b the gap extension penalty. However, analyses of alignment of homologous sequences have shown, both for protein and nucleic sequences, that this model under-estimates the probability of long indels (7, 13, 14). Indeed, more realistic indel penalties can be estimated with models of the following form:

$$p = a + b\log(L)$$

However, because of computational complexity, such models have not been implemented in commonly used alignment programs. Fortunately, other approaches have been proposed to align sequences with large indels (see *Section 4.3*).

The probability of occurrence of indels in proteins also depend on the degree of divergence between sequences (7, 13). Thus, as for amino-acids substitution matrices, indel penalty parameters should ideally be varied according to the divergence of the sequences to be aligned. The probability also depends on the nature of the sequences: protein, structural RNA, non-coding DNA (in which transposable elements may be inserted), etc. Moreover, probabilities of indel may vary along sequences. In proteins notably, indels are more frequent within external loops than in the core of the structure. Thus, knowledge on the structure of proteins can be used to weight indels. For example, the CLUSTAL W program uses residue specific indel penalties and locally reduced indel penalties to encourage new gaps in potential loop regions rather than in regular secondary structure. In cases where secondary structure information is available, indel-penalty masks can also be used to guide the alignment.

It is important to note that, in most programs, default parameters for gap penalties have been set for typical globular proteins. These may not be optimal for other sequences.

3 Searching for homologous sequences

The first step in the analysis of a family of homologous sequences consists of searching for all available members of that family. Published sequences are stored in databases: GenBank (15) or EMBL (16) for nucleic acid sequences and SWISSPROT-TREMBL (17) or PIR (18) for protein sequences. Retrieval systems such as Entrez (19), SRS (20), or ACNUC (21) have been developed to query those databases and extract sequences according to the associated annotation (e.g. keywords, taxon, authors). Some WWW addresses for commonly used database retrieval systems are shown in *Table 2*. Unfortunately, it is not possible to rely on the annotation to identify in a database all homologous sequences belonging to a given family. Presently, the most efficient way to identify those homologues

Table 2 Websites for text-based searches in sequence databases

Entrez at NCBI	http://www.ncbi.nlm.nih.gov/Entrez/
SRS at EBI	http://srs.ebi.ac.uk/
WWW-QUERY at PBIL	http://pbil.univ-lyon1.fr/
ExPASy	http://www.expasy.ch/sprot/
DBGET at GenomeNet	http://www.genome.ad.jp/

Table 3 Websites for sequence similarity searches in databases

BLAST at NCBI[a]	http://www.ncbi.nlm.nih.gov/BLAST/
WU-BLAST at EMBL[b]	http://dove.embl-heidelberg.de/Blast2e/
FASTA at EBI	http://www2.ebi.ac.uk/fasta3/
Smith-Waterman search at EBI	http://www2.ebi.ac.uk/bic_sw/
BCM search launcher	http://gc.bcm.tmc.edu:8088/search-launcher.html
BLAST at PBIL[c]	http://pbil.univ-lyon1.fr/BLAST/blast.html

[a] Possibility to select BLAST output results by taxa.

[b] Performs multiple alignment on homologous sequences detected by BLAST.

[c] Possibility to select BLAST output results by taxa or keyword.

consists in taking one member of the family and comparing it to the entire database with a similarity search program such as FASTA (22) or BLAST (23, 24). To guarantee a more exhaustive search, one may repeat this procedure with several distantly related homologues identified in the first step. See the review by Altschul, *et al.* (25) for a comprehensive discussion of sequence similarity searches.

The sensitivity of a sequence similarity search may be improved by weighting sites according to their degree of conservation. Thus, once several homologous sequences have been identified, it is possible to use methods such as profile searches (see Chapter 5 in this volume) or PSI-BLAST (24) that rely on a multiple alignment to identify more distantly related members of the family. A list of some similarity search WWW servers is presented *Table 3*.

4 **Multiple alignment methods**

Once homologous sequences have been identified, which program should be preferentially used to align them? Several multiple alignment methods (algorithms) have been developed, but none of them is ideal. Thus, it is important to have an idea of what these algorithms try to solve, in order to make an informed choice of the most appropriate method(s) for a particular problem. The multiple alignment problem is algorithmically hard: methods that guarantee to find the best alignment (for a given measure of alignment score and for a given set of substitution matrix and gap penalty parameters) require so much time and space (memory) that they cannot be used in practice with, say, more than 10 to 15 sequences of length 100. Thus, alternative algorithms have been developed using heuristics to gain speed and limit space requirements. Although these heuristics do not guarantee to find the optimal alignment, they are very useful in practice and often give results very close to the exact solution. In the following we will focus on four families of multiple alignment algorithms:

(a) Algorithms that guarantee to find the optimal alignment for a given scoring scheme; these algorithms can be used only for a limited number of short sequences.

(b) Heuristic algorithms that are based on a progressive pairwise alignment approach.

(c) Heuristic algorithms that build a global alignment based on local alignments.

(d) Heuristic algorithms that build local multiple alignments.

It should be noted that this list is not exhaustive. Other multiple alignment methods such as those based on hidden Markov models (26) or genetic algorithms

Table 4 Websites for multiple alignments

• Optimal global multiple alignment	
MSA at IBC	http://www.ibc.wustl.edu/ibc/msa.html
• Progressive global multiple alignment	
ClustalW at EBI[a]	http://www2.ebi.ac.uk/clustalw/
ClustalW, Multalin at PBIL[b]	http://pbil.univ-lyon1.fr/
MAP, ClustalW at BCM	http://kiwi.imgen.bcm.tmc.edu:8088/search-launcher/launcher.html
Multalin at INRA[b]	http://www.toulouse.inra.fr/multalin.html
ClustalW at Pasteur[c]	http://bioweb.pasteur.fr/seqanal/alignment/intro-uk.html
ClustalW at DDBJ	http://www.ddbj.nig.ac.jp/searches-e.html
MAP	http://genome.cs.mtu.edu/map.html
• Block-based global multiple alignment	
DCA at BiBiServ	http://bibiserv.techfak.uni-bielefeld.de/dca/
DIALIGN2 at BiBiServ	http://bibiserv.TechFak.Uni-Bielefeld.DE/dialign/
DCA at Pasteur[c]	http://bioweb.pasteur.fr/seqanal/alignment/intro-uk.html
DIALIGN2 at Pasteur[c]	http://bioweb.pasteur.fr/seqanal/alignment/intro-uk.html
ITERALIGN at Stanford	http://giotto.stanford.edu/~luciano/iteralign.html
• Motif-based local multiple alignment	
MEME at SDSC	http://www.sdsc.edu/MEME/
MEME at Pasteur	http://bioweb.pasteur.fr/seqanal/motif/meme/
MATCH-BOX	http://www.fundp.ac.be/sciences/biologie/bms/matchbox_submit.html
BLOCK Maker at FHCRC	http://www.blocks.fhcrc.org/blockmkr/make_blocks.html
PIMA at BCM	http://kiwi.imgen.bcm.tmc.edu:8088/search-launcher/launcher.html
PIMA II at BMERC	http://bmerc-www.bu.edu/protein-seq/pimall-new.html

[a] Possibility to display and edit alignment with the JALVIEW JAVA applet.

[b] Coloured alignments.

[c] In combination with many WWW tools for molecular phylogeny.

Table 5 Software for multiple alignments

ClustalW (UMPV)[a]	ftp://ftp-igbmc.u-strasbg.fr/pub/ClustalW/
ClustalX (UMPV)[a] (ClustalW + graphical interface)	ftp://ftp-igbmc.u-strasbg.fr/pub/ClustalX/
Multalin	http://www.toulouse.inra.fr/multalin.html
MSA (U)[a]	http://www.ibc.wustl.edu/ibc/msa.html
DIALIGN (U)[a]	http://www.gsf.de/biodv/dialign.html
DCA (U)[a]	http://bibiserv.techfak.uni-bielefeld.de/dca/
RIW/DNR (U)[a]	ftp://ftp.genome.ad.jp/pub/genome/saitama-cc/
MACAW (MP)[a]	ftp://ftp.bio.indiana.edu/molbio/align/macaw/

[a] Availability: U = UNIX , M = Macintosh, P = PC, V = VMS.

(27) can also be used. For a review of multiple alignment algorithms see reference (28).

Many of the programs reviewed here can be used directly through the WWW (see *Table 4*) or downloaded over the Internet to be installed on a local computer (see *Table 5*).

4.1 Optimal methods for global multiple alignments

In this section, we will mention several methods that are said to be optimal, because they guarantee to find the 'best' multiple alignment among all possible solutions for a given scoring scheme. It should be stressed that the term 'optimal' is taken here in its mathematical meaning. Whether a mathematically optimal alignment corresponds or not to the biologically correct alignment (i.e. the alignment that represent the most likely evolutionary scenario) will depend on the choice of parameters (weighting of substitutions and of indels, see *Section 2.3*) and on the way the multiple alignment is scored.

4.1.1 Scoring schemes for multiple alignments

In principle, the score of a multiple alignment should reflect its likelihood (according to a given evolutionary model). There are different ways to measure the score (or cost) of a multiple alignment. In the following we consider that a sequence is an ordered set of letters taken from an alphabet Σ. An alignment of n sequences $S1, \ldots, Sn$ can be defined as a matrix $a(S1, \ldots ,Sn) = A$, where each entry Aij is either a letter from Σ or a null symbol (the gap symbol, usually denoted by –). The row i from A is the sequence S_i, after gaps are removed.

In the simplest model, the cost of an alignment of n sequences is defined as the sum of the cost of its columns. However, this model is crude because each column of the alignment is considered independently of its context (i.e. a gap of length L is considered as corresponding to L independent indels).

In more realistic models, a gap is interpreted as one single mutational event (a deletion or an insertion of L residues) and associated with a cost that depends on its length (see *Section 2.3.2*). With such models, pairwise alignment costs are defined as the sum of substitution and gap costs. However, the definition of the multiple alignment cost is more complex. One possible solution, known as the Sum of Pairs (SP) alignment cost (29), consists of calculating the multiple alignment cost from pairwise alignment costs. A multiple alignment $a(S1, \ldots, Sn)$ contains $n(n-1)/2$ pairwise alignments $a(Si.,Sj)$ where $1 \leq i < j \leq n$. Each *projection* $a(Si.,Sj)$ is the pairwise alignment built from $a(S1, \ldots, Sn)$ by removing all the rows except the rows i and j, and then by removing all the columns that contains two null letters. The SP multiple alignment cost is defined as the sum of all its projections costs (29).

Simple SP alignment cost may, however, be inappropriate when some groups of sequences are heavily over or under-represented in a family. This drawback may be corrected by introducing a proper weighting system (30, 31) which assigns a weight to each sequence. This can be used to give less weight to sequences from overrepresented groups. Another solution consists of using a cost function based on an evolutionary tree. The tree leaves are the sequences we want to align, and the internal nodes are their hypothetical ancestral sequences. For a given tree, the cost of an alignment $a(S1, \ldots, Sn)$ is the sum of all its projections $a(Si.,Sj)$ on adjacent sequences Si and Sj in the tree (32).

4.1.2 Algorithmic complexity of optimal multiple alignment methods

The optimal alignment is the one with the maximal score (or the minimal cost). Needleman and Wunsh (33) proposed an efficient algorithm, based on dynamic programming, to compute this minimal cost for pairwise alignments. This dynamic programming approach can be easily generalized to more than two sequences. However, computing the minimal alignment cost of n sequences, each of length l requires $o(2^n l^n)$ time and $o(l^n)$ space (i.e. time proportional to $2^n l^n$ and computer memory proportional to l^n) and the complexity is even higher if gap cost are not linear (see *Section 2.3.2*). Such an algorithm cannot be used, in practice, for much more than three sequences. For example, to align ten sequences of length 100, on a very fast computer that would need 10^{-9} sec to compute the score for one column of a multiple alignment, it would take approximately three million years ($2^{10} \, 100^{10} \, 10^{-9} \cong 10^{14}$ sec) to compute the alignment. This assumes we have approximately ten billion giga-bytes of memory.

Carrillo and Lipman (29) proposed a branch and bound algorithm to compute a minimal SP cost alignment. This algorithm uses an upper bound of the alignment SP cost to limit the space and time used by dynamic programming. This approach is implemented in the program MSA (34). A new version of MSA with substantial improvements in time and space usage is available (35). Despite these improvements, MSA cannot easily be used for more than about 10 short sequences.

As stated previously, cost functions based on an evolutionary tree are, in principle, better than SP alignment costs to measure the likelihood of an alignment. However, the alignment problem under an evolutionary tree is even harder than the SP alignment problem, as the algorithm has to find the alignment, the tree, and the ancestral sequences such that the alignment cost is minimal. The problem remains hard even if the tree is given (36).

4.2 Progressive global alignment

Progressive alignment is the most commonly used method to align biological sequences. This heuristic approach is very rapid, requires low memory space and offers good performance on relatively well-conserved, homologous sequences (37, 38).

4.2.1 Description of progressive alignment methods

Progressive alignment consists of building a multiple alignment using pairwise alignments in three steps:

(a) Compute the alignment scores (or distances) between all pairs of sequences.

(b) Build a *guide* tree that reflects the similarities between sequences, using the pairwise alignment distances.

(c) Align the sequences following the guide tree. Corresponding to each node in the tree, the algorithm aligns the two sequences or alignments that are associated with its two daughter nodes. The process is repeated beginning from the tree leaves (the sequences) and ending with the tree root.

Depending on the algorithms, steps (b) and (c) are done separately, or merged in one step where the tree topology is deduced from the progressive alignment process.

Figure 2 illustrates the progressive alignment process. *S1* is first aligned with *S2* following the given tree, *S3* is then aligned with *S4*, then the two alignments α*(S1,S2)* and α*(S3,S4)* are aligned together, and finally *S5* is aligned with α*(S1,S2,S3,S4)*. Notice that even if α*(S1,S2)* and α*(S3,S4)* are optimal alignments computed by dynamic programming, the progressive alignment approach does not guarantee that α*(S1,S2,S3,S4)* is optimal for a multiple alignment cost function (SP cost or the tree cost for example).

A great number of tools that use a progressive alignment approach have been proposed, they differ by the methods used in at least one of the three steps.

In the first step (a) the pairwise alignment cost can be computed by dynamic programming, or by heuristic algorithms. The multiple alignment program CLUSTAL W (9) for example allows one to choose either dynamic programming or a heuristic method. Dynamic programming gives more accurate scores but is slower than heuristic methods.

Different algorithms can be used to build a tree (step b) given a distance matrix between sequences. Following Feng and Doolittle (37), early versions of CLUSTAL (39) used the UPGMA algorithm (40). However, UPGMA is notorious for giving incorrect branching orders when rates of substitution vary in different lineages. Therefore, CLUSTAL W (9) now uses the Neighbor-Joining (41) algorithm to build the guide tree.

The main problem in the third step (c) consists of aligning two alignments.

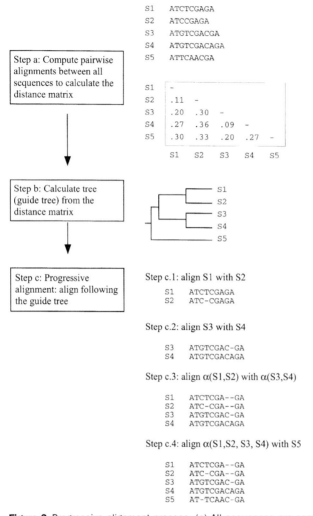

Figure 2 Progressive alignment process. (a) All sequences are compared to each other S_2. (b) A guide tree is calculated from the pairwise distance matrix. (c) Sequences are progressively aligned following the guide tree.

The simplest method for this problem reduces each alignment to a *consensus* sequence, and uses a pairwise alignment algorithm to do the work. In the consensus sequence, each column of the alignment is represented by its most frequent letter. Consensus alignment was used in the first version of CLUSTAL. In most programs, each alignment is considered as a *profile* (see Chapter 5). In a profile, a column is reduced to a distribution giving the frequency of each letter. Two profiles are aligned as two sequences by dynamic programming without major modification of the algorithm. The alignment of two profiles of length l takes $o(a^2l^2)$, where a is the alphabet size. CLUSTAL W uses profile alignment with position-specific gap penalties (see *Section 2.3.2*).

4.2.2 Problems with progressive alignment methods

An important problem with this progressive alignment approach stems from the 'greedy' nature of the algorithm: any mistakes that appear during early alignments cannot be corrected later as new sequence information is added. For example, suppose that we have to align three sequences (x, y, z). Consider a short fragment of these sequences for which the optimal alignment is:

```
x  ACTTA
y  A-GTA
z  ACGTA
```

Suppose that the guide tree based on pairwise comparison of entire sequences indicates that we should first align sequence x with sequence y, followed by the alignment of sequence z with the first two (already aligned together). At the first step, there are three possible alignments of x and y giving exactly the same score:

```
x  ACTTA     x  ACTTA     x  ACTTA
y  A-GTA     y  AGT-A     y  AG-TA
```

At the later step, the gap that was introduced cannot be changed. Thus adding sequence z could give the following three alignments:

```
x  ACTTA     x  ACTTA     x  ACTTA
y  A-GTA     y  AGT-A     y  AG-TA
z  ACGTA     z  ACGTA     z  ACGTA
```

Only the first of these alignments is optimal. At the first step, only one of the three possibilities will be used. If it is the wrong one, we cannot correct this later.

To avoid that problem, iterative optimization strategies such as RIW or DNR (42) have been proposed. These methods are reported to perform better than CLUSTAL W (42). However, although these methods are much faster than optimal algorithms, they are still to slow for large dataset.

Another limitation of the progressive approach described above is that it requires computing pairwise distances between all sequences to calculate the guide tree. One may sometimes have to align set of homologous sequences that include some non-overlapping fragments (e.g. partial protein sequences). When sequences are non-overlapping they are obviously completely unrelated and thus the guide tree generated may be totally false. The alignment produced in this case can be unpredictable.

4.3 Block-based global alignment

The sequences to be compared may share conserved blocks, separated by non-conserved regions containing large indels. In such cases, the result of optimal or progressive global alignment methods will depend greatly on the choice of gap penalty parameters. An alternative to these approaches consists of searching for

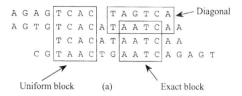

Uniform block   (a)   Exact block

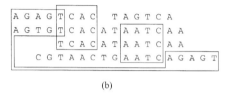

(b)

Figure 3 (a) Consistent set of blocks. (b) Non-consistent set of blocks.

conserved blocks that will be used as anchors in order to align the sequences. Blocks are alignments of fragments (segments) of sequences (local alignments). Most methods consider gap-free blocks. Depending on the programs used, the blocks allowed can be *exact* (composed of identical segments) or not exact and they may be *uniform* (found in every sequence) or not. The selected set of blocks must be *consistent*, i.e. the blocks can occur together in a multiple global alignment (*Figure 3*). Once blocks have been computed, it is possible to use a classical approach to align regions between blocks (e.g. ref. 43).

The first multiple block alignment program (44) used a sorting algorithm in order to compute uniform exact blocks. Faster algorithms based on suffix trees (45), or equivalent data structures, can also be used to compute exact blocks. However, homologous regions are rarely exactly conserved. ASSEMBLE (46) performs a dot matrix analysis on all pairs of sequences and then compares these dot matrices to find uniform blocks that are not necessarily exact. In practice, it often happens that some blocks are not present in all sequences. Thus, a further improvement has consisted of developing methods that allow blocks that are not necessarily uniform. DIALIGN (47, 48) is based on computing gap-free blocks between pairs of segments (*diagonals*).

A set of uniform blocks is consistent when each pair of blocks is ordered (they do not cross each other). Using this observation, selecting an optimal consistent set of blocks can be reduced to a classic optimal-path algorithm in a graph (44). The optimal-path algorithm requires $o(M^2)$ time for M blocks. Faster algorithms (sub-quadratic) have been proposed in order to compute an optimal consistent uniform set of blocks (49, 50). However, finding an optimal consistent set of *non-uniform* blocks is an intractable problem (51). Indeed, the consistency of non-uniform blocks cannot be reduced to a binary relation between them. A set of three non-uniform blocks, such that all its three pairs of blocks are consistent, is not necessarily consistent. To compute a 'good' consistent set of diagonals, DIALIGN uses a heuristic algorithm in which diagonals are incorporated by decreasing score order into a consistent set of diagonals. Diagonals not consistent with the set of selected diagonals are rejected. In order to check if a new diagonal is consistent or not with the set of selected diagonals, DIALIGN maintains a data structure in $o(kL^2)$ time, for k sequences of total length L. This makes it slower than progressive alignment programs. This computation time can however be reduced to $o(k^2L + L^2)$ (52) and even, thanks to recent developments, to $o(L^2)$. Thus, faster versions of block-based alignment methods should be available in the near future.

4.4 Motif-based local multiple alignments

The sequences to compare may share similar regions, without necessarily being globally related. These homologous modules may occur in different relative positions and may be duplicated in different sequences. In such cases it is not possible to compute a global alignment, but one may look for 'good' local alignments of segments taken in the sequences. Calculating local alignments consists of finding approximate repeated patterns in a set of sequences. Dynamic programming has been adapted in order to find the maximal diagonal score for pairwise comparison (53). For more than two sequences the problem is hard and heuristics are needed as for the global multiple alignment problem.

PRALIGN (54) computes consensus words for a given word length. For each possible word w of length k one may define the *neighbourhood* of w as the set of k length words whose score with w is higher than a given cut-off. The score of w is then the sum of all the scores with his neighbours that occur in the given sequences. PRALIGN tries to compute the best score words (consensus words) of fixed length. The main problem with this program is its space requirement: for a fixed length k the space used is proportional to 20^k (for proteins). This space requirement could be much reduced using automatons as it is done in BLAST.

The MACAW method (55) combines pairwise comparisons in order to compute multiple local alignments. In a first step, MACAW marks, for each pair of sequences, all the diagonals with significant scores. The diagonals are then merged into local alignments. MACAW is generally considered too time-consuming for a general local alignment method, as it needs $o(L^2)$ time for the first step (L is the sum of the sequence lengths).

Most recent local alignment programs are based on statistical methods. Statistical methods use computationally efficient heuristics in order to solve optimization problems. GIBBS (56) uses iterative Gibbs sampling in order to find blocks. The computation time of this approach grows linearly with the number of input sequences. GIBBS is available in the programs MACAW and Block Maker (57). The tool MEME (58) uses an expectation-maximization (EM) algorithm (59, 60) to locate repeated patterns.

4.5 Comparison of different methods

When sequences are similar (say more that 50% pairwise identity for proteins, 70% for DNA) and are homologous over their entire length, all global alignment methods give more or less correct results. Moreover, in such cases, any reasonable set of parameters (substitution matrix, and usually, gap opening and gap extension penalties) will give similar alignments. However, when at least two sequences in a given family share less identity, or if homologous regions are interrupted by large gaps of different sizes, the result of alignment may vary considerably according to programs and parameters used.

Several comparative analyses of multiple alignment programs have been published (42, 48, 61, 62). These comparisons are based on the ability to detect motif patterns on several protein families or based on reference alignments derived from three-dimensional protein structures. Comparative analysis can also be based on the effect of the multiple alignment programs on phylogeny. Such a study was done on 18S rDNA from 43 protozoan taxa (63). These comparisons must be taken only as indications. Indeed, the parameter values (substitution matrices, gap penalty, etc.) used in these comparisons may not be optimal for other sequence families (61). In addition these parameters are not really comparable, even if the programs use the same strategies. For example a gap opening score of 5 does not have the same meaning in CLUSTAL W (9) as it does in MULTAL (38), as the value 5 will be modified in the programs (multiplied by constants for example). For these reasons and because no known method guarantees to find the *correct* alignment, it is still necessary to combine different methods from different families of algorithms and human expertise to obtain satisfactory alignments.

Figure 4 summarizes indications to guide users in their choice according to the sequences they have to align. For the alignment of two sequences, one should use an optimal pairwise alignment method (for example LALIGN or SIM (64), see *Table 6*). For more than two sequences, one generally has to use heuristic approaches. As a first step, the user should try to compute the multiple alignment with a progressive alignment program. These programs are rapid, do not demand large memory capacity and may thus be run on large dataset even on micro-computers. Among programs using this approach, we recommend CLUSTAL W (or its graphical user interface version: CLUSTAL X) (65, 66). This includes useful features such as automatic selection of amino-acid substitution matrix during alignment and lower weighting of gaps in potential protein loops. If this first alignment shows that all sequences are related to each other over their entire lengths, it is unlikely that any other method will give a better result (*Figure 4a*).

However, if there are some highly divergent sequences, large gaps, or poorly conserved regions it is recommended to compare the results of different methods and/or sets of parameters. *Figure 4b* shows homologous sequences sharing conserved blocks separated by non-conserved regions of varying size. This situation, which is frequently observed in practice (e.g. in genomic DNA sequences and in many protein families), is particularly error prone for progressive alignment methods, notably because the linear weighting of gaps tends to over-penalize long indels. Block-based global methods (e.g. DIALIGN, ITERALIGN) (47, 48, 67) are not sensitive to these long gaps and are particularly appropriate for such

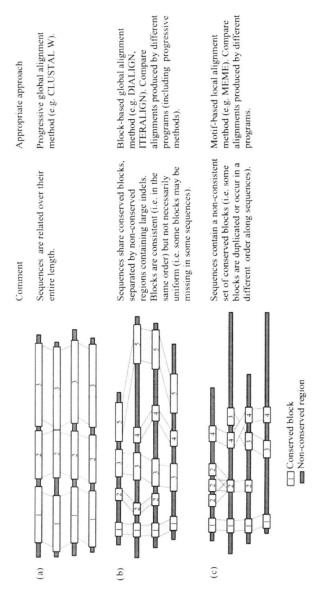

Figure 4 Choice of multiple alignment methods according of the nature of the sequence set.

67

Table 6 Websites for pairwise alignments

LFASTA at PBIL[a]	http://pbil.univ-lyon1.fr/lfasta.html
SIM at ExPASy[a]	http://www.expasy.ch/sprot/sim-prot.html
BLAST two sequences at NCBI	http://www.ncbi.nlm.nih.gov/gorf/bl2.html
LALIGN at CRBM	http://www2.igh.cnrs.fr/bin/lalign-guess.cgi
SIM, GAP, NAP, LAP	http://genome.cs.mtu.edu/align/align.html

[a] Possibility to visualize pairwise alignments with LALNVIEW (83).

cases. Moreover, one drawback of progressive methods (but also of optimal global alignment methods) is that an alignment is produced even if sequences are not related, possibly of random origin. DIALIGN and ITERALIGN on the contrary do not attempt to generate a global alignment if sequences are only locally related. Another interesting feature of these programs is that they indicate the significance of the alignment: in DIALIGN for example, regions that are not considered to be aligned (e.g. a non-conserved region between two aligned blocks) are printed as lower-case letters whereas aligned residues are in upper-case.

Global methods (optimal, progressive, or block-based) are appropriate only if all conserved blocks are consistent (see *Figure 3*). If, as presented in *Figure 4c*, some domains are duplicated, or ordered differently along sequences it is necessary to use a local multiple alignment method to align all related domains. The WWW version of the MEME tool (see *Table 4*) provides a graphical representation of the motifs found in sequences which proves to be very helpful to analyse the domain organization of proteins.

4.6 Particular case: aligning protein-coding DNA sequences

It is sometimes necessary to align protein-coding DNA sequences rather than proteins. Two examples are the design of primers to identify related genes by PCR or for molecular phylogenies relying on the measure of substitution rates at synonymous (Ks) or non-synonymous (Ka) sites of codons. Due to the degeneracy of the genetic code, it is generally more difficult to align coding DNA sequences than their protein translation. Moreover, some ambiguities in DNA alignments may be solved when considering the protein translation. For example, the two DNA alignments below have exactly the same similarity score:

```
      L    F              L    F
     CTT  TTC            CTT  TTC
     CTC  ---            ---  CTC
      L    -              -    L
    (a)                (b)
```

However, the second alignment can be rejected unambiguously taking into account the protein translation. Thus, the procedure commonly used to align protein-coding DNA sequences is the following:

(a) Extract coding DNA sequences and the corresponding protein translation.

(b) Align protein sequences.

(c) Back-translate the protein alignment into a nucleic acid alignment.

The program PROTAL2DNA (C. Letondal, unpublished) has been written for this purpose, and is available at the Pasteur WWW server:

http://bioweb.pasteur.fr/seqanal/interfaces/protal2dna.html

Note that the WWW-QUERY server (see *Table 2*) may be used to extract both coding DNA sequences and their corresponding protein translation (taking into account species- or organelle-specific genetic codes).

5 Visualizing and editing multiple alignments

Results of multiple alignment programs are generally saved simply as text files. There is presently no standard format for multiple alignments. However, the MSF output format (*Figure 5*) is provided by most of popular alignment programs and is recognized by many programs that require alignments as an input (e.g. molecular phylogeny, profile searches). The MASE format presents the advantage of allowing the inclusion of annotation regarding the whole alignment or specific to each sequence (*Figure 6*). Textual representation of multiple alignments is,

```
PileUp


   MSF:   171  Type: P    Check:  8689   ..

  Name: BTG1_BOVIN      oo  Len:  171  Check:  4676  Weight:  1.00
  Name: BTG1_CHICK      oo  Len:  171  Check:  3006  Weight:  1.00
  Name: BTG2_HUMAN      oo  Len:  171  Check:  5090  Weight:  1.00
  Name: BTG2_MOUSE      oo  Len:  171  Check:  5917  Weight:  1.00

//


BTG1_BOVIN      MHPFYSRAAT MIGEIAAAVS FISKFLRTKG LTSERQLQTF SQSLQELLAE
BTG1_CHICK      MHPALYTRAS MIREIAAAVA FISKFLRTKG LMNERQLQTF SQSLQELLAE
BTG2_HUMAN      ..MSHGKGTD MLPEIAAAVG FLSSLLRTRG CVSEQRLKVF SGALQEALTE
BTG2_MOUSE      ..MSHGKRTD MLPEIAAAVG FLSSLLRTRG CVSEQRLKVF SRALQDALTD


BTG1_BOVIN      HYKHHWFPEK PCKGSGYRCI RINHKMDPLI GQAAQRIGLS SQELFRLLPS
BTG1_CHICK      HYKHHWFPEK PCKGSGYRCI RINHKMDPLI GQAAQRIGLS SQELFQLLPS
BTG2_HUMAN      HYKHHWFPEK PSKGSGYRCI RINHKMDPII SRVASQIGLS QPQLHQLLPS
BTG2_MOUSE      HYKHHWFPEK PSKGSGYRCI RINHKMDPII SKVASQIGLS QPQLHRLLPS


BTG1_BOVIN      ELTLWVDPYE VSYRIGEDGS ICVLYEASPA GGSTQNSTNV QMVDSRISCK
BTG1_CHICK      ELTLWVDPYE VSYRIGEDGS ICVLYEAAPA GGS.QNNTNM QMVDSRISCK
BTG2_HUMAN      ELTLWVDPYE VSYRIGEDGS ICVLYEEAPL AAS....... ...CGLLTCK
BTG2_MOUSE      ELTLWVDPYE VSYRIGEDGS ICVLYEEAPV AAS....... ...YGLLTCK


BTG1_BOVIN      EELLLGRTSP SKNYNMMTVS G
BTG1_CHICK      EELLLGRTSP SKSYNMMTVS G
BTG2_HUMAN      NQVLLGRSSP SKNYVMAVSS .
BTG2_MOUSE      NQMMLGRSSP SKNYVMAVSS .
```

Figure 5 Example of alignment in MSF format.

```
;; Header (at least one line)
; Sequence-specific annotation  (at least one line)
sequence name 1
ATA-GGGA ....
; Sequence-specific annotation  (at least one line)
sequence name 2
ATA-GGGA ....
; Sequence-specific annotation  (at least one line)
sequence name 3
ATAGGGGA ....
etc
```

Example:

```
;;Aligned by clustal on Wed Dec  2 09:12:04 1998
;;Block: conserved domain: 49..89
;;Group of species = 2 BTG1: 1, 2
;AC   P53348; O18950;
;DE   BTG1 PROTEIN (B-CELL TRANSLOCATION GENE 1 PROTEIN) (MYOCARDIAL
;DE   VASCULAR INHIBITION FACTOR) (VIF).
BTG1_BOVIN
MHPFYSRAATMIGEIAAAVSFISKFLRTKGLTSERQLQTFSQSLQELLAEHYKHHWFPEK
PCKGSGYRCIRINHKMDPLIGQAAQRIGLSSQELFRLLPSELTLWVDPYEVSYRIGEDGS
ICVLYEASPAGGSTQNSTNVQMVDSRISCKEELLLGRTSPSKNYNMMTVSG
;AC   P34743;
;DE   BTG1 PROTEIN (B-CELL TRANSLOCATION GENE 1 PROTEIN).
BTG1_CHICK
MHPALYTRASMIREIAAAVAFISKFLRTKGLMNERQLQTFSQSLQELLAEHYKHHWFPEK
PCKGSGYRCIRINHKMDPLIGQAAQRIGLSSQELFQLLPSELTLWVDPYEVSYRIGEDGS
ICVLYEAAPAGGS-QNNTNMQMVDSRISCKEELLLGRTSPSKSYNMMTVSG
;AC   P78543;
;DE   BTG2 PROTEIN PRECURSOR (NGF-INDUCIBLE ANTI-PROLIFERATIVE
;DE    PROTEIN PC3).
BTG2_HUMAN
--MSHGKGTDMLPEIAAAVGFLSSLLRTRGCVSEQRLKVFSGALQEALTEHYKHHWFPEK
PSKGSGYRCIRINHKMDPIISRVASQIGLSQPQLHQLLPSELTLWVDPYEVSYRIGEDGS
ICVLYEEAPLAAS---------CGLLTCKNQVLLGRSSPSKNYVMAVSS-
;AC   Q04211;
;DE   BTG2 PROTEIN PRECURSOR (NGF-INDUCIBLE PROTEIN TIS21).
BTG2_MOUSE
--MSHGKRTDMLPEIAAAVGFLSSLLRTRGCVSEQRLKVFSRALQDALTDHYKHHWFPEK
PSKGSGYRCIRINHKMDPIISKVASQIGLSQPQLHRLLPSELTLWVDPYEVSYRIGEDGS
ICVLYEEAPVAAS---------YGLLTCKNQMMLGRSSPSKNYVMAVSS-
```

Figure 6 MASE format. This format is used to store nucleotide or protein multiple alignments along with annotations relative to the whole alignment (indicated in the header), or specific to each sequence. The beginning of the file must contain a header containing at least one line (but the content of this header may be empty). The header lines begin by ';;'. The body of the file has the following structure: First, each entry begins with one (or more) annotation lines. Annotation lines begin by the character ';'. Again, this annotation line may be empty. After the annotations, the name of the sequence is written on a separate line. At last, the sequence itself is written on the following lines.

however, poorly informative. Therefore, graphical interfaces have been developed to manipulate and edit multiple alignments. Generally, these interfaces allow users to colour or shade residues (amino acids or nucleotides) according to various criteria such as physico-chemical properties, degree of conservation within the alignment, hydrophobicity or secondary structure. The use of colours is very helpful to interpret a multiple alignment. It gives a much more comprehensive view of the information embedded in a multiple alignment than a simple textual representation. Besides, these interfaces propose several interesting facilities detailed below. A list of such graphical interfaces is given in *Table 7*.

Table 7 Multiple alignment viewers and editors

Jalview (J)[ab]	http://www2.ebi.ac.uk/~michele/jalview/contents.html
CINEMA 2.1 (J)[ac]	http://www.biochem.ucl.ac.uk/bsm/dbbrowser/CINEMA2.1/
SEAVIEW (U)[a]	http://pbil.univ-lyon1.fr/software/seaview.html
MPSA (UM)[a]	http://www.ibcp.fr/mpsa/
Se-Al (M)[a]	http://evolve.zps.ox.ac.uk/Se-Al/Se-Al.html
ClustalX (UMPV)[a] (ClustalW + graphical interface)	ftp://ftp-igbmc.u-strasbg.fr/pub/ClustalX/
DCSE (U)[a]	http://indigo2.uia.ac.be:80/~peter/dcse/

[a] Availability: U = UNIX , M = Macintosh, P = PC, V = VMS, J = JAVA applet.

[b] Links to sequence databases.

[c] Possibility to download alignments from the PRINTS database.

5.1 Manual expertise to check or refine alignments

Whatever the quality of the software, it is necessary to examine the alignment to check that there are no obvious errors. Alignments of sequences with large length differences, or with duplicated domains are particularly error prone, even if the sequences are not very divergent. A good control consists in verifying that local similarities detected by pairwise sequence comparisons are preserved in the multiple alignment. For such purposes, one may use the results of similarity searches (BLAST, FASTA, etc.), or run pairwise local alignment software (see *Table 6*) or use a dot-plot representation. Pairwise alignments can be computed directly from JALVIEW. SEAVIEW (68) includes a dot-plot utility that can be used to drive the alignment, semi-automatically. CINEMA (69) allows user to run a BLAST search or a dot-plot on selected sequences.

In some cases, it may be necessary to refine part of the alignment. Experienced users are often able to recognize residues that have been misaligned. In some cases, external information (e.g. known interactions or 3-D structures) may also reveal alignment errors. SEAVIEW and CLUSTALX allow users to run CLUSTALW on a specified region and/or a specified set of sequences, without changing the rest of the alignment.

Alignment editors (except CLUSTALX) also allow users to manually add or remove gaps in the alignment. In some interfaces (e.g. JALVIEW or SEAVIEW), it is possible to define groups in order to edit a subset of sequences. In the absence of objective criteria, manual alignment editing should, however, be used with caution.

5.2 Annotating alignments, extracting sub-alignments

The SEAVIEW software allows users to annotate alignments (e.g. to indicate the location of relevant features such as enzyme active sites or RNA splicing signals). The locations of annotations are correctly preserved after indels are inserted or moved. This software also allows one to define groups of sequences and blocks in the alignment and thus to extract sub-alignments. This feature is particularly useful when building phylogenetic trees where one needs to exclude unreliable parts of alignments (i.e. regions for which the alignment is ambiguous). It is also useful to select particular domains for profile searches. Definitions of groups and blocks can be saved along with the alignment in MASE format (*Figure 6*).

5.3 Comparison of alignment editors

Each of the editors presented in *Table 7* has some specific useful features, some of which have been mentioned above. Programs written in JAVA (JALVIEW, CINEMA) present two advantages. First, they can be used from any computer and run directly from a WWW browser (although, depending on the network load, the time necessary to download the JAVA applet through the internet sometimes limits considerably their usefulness). Secondly, thanks to the network communication facilities provided by JAVA, these programs allow users to directly access information stored in sequence databases available on the internet. CLUSTALX is a graphical interface to the CLUSTALW program and not simply an alignment viewer. However, it does not allow manual editing of alignments. MPSA is dedicated to protein secondary structure prediction. SEAVIEW is particularly suited for phylogenetic analyses and can notably be used in combination with the PHYLOWIN graphical interface dedicated to molecular phylogeny (68).

5.4 Alignment shading software, pretty printing, logos, etc.

To publish the results of such analyses, it is generally useful to prepare a high quality colour figure of the multiple alignment. Some of the above editors (e.g. JALVIEW, CINEMA) can be used to save or print coloured alignments in a format suitable for publication. Other programs, some of which are available on the WWW, have been developed specifically for that purpose (see *Table 8*). The program LOGO (70) is intended to give a visual representation of a consensus sequence, along with possible variants.

6 Databases of multiple alignments

Databases of precompiled multiple alignments have been developed, essentially for protein sequences (71–79) but also for rRNA (80–82) and some other nucleic acid sequences (see *Table 9*). The approach used to cluster together homologous protein sequences varies according to databases. Some intend to classify together proteins homologous over their entire length (protein families), whereas others focus on the classification of protein domains (see *Table 9*). For example,

Table 8 Pretty printing, shading, logos, etc.

BOXSHADE	http://ulrec3.unil.ch/software/BOX_form.html
WebLogo	http://www.bio.cam.ac.uk/cgi-bin/seqlogo/logo.cgi
Mview	http://mathbio.nimr.mrc.ac.uk/nbrown/mview/
AMAS	http://barton.ebi.ac.uk/servers/amas_server.html

Table 9 Databases of multiple alignments

• Protein families	
PIRALN	http://www-nbrf.georgetown.edu/nbrf/getaln.html
HOVERGEN	http://pbil.univ-lyon1.fr/databases/hovergen.html
PROTOMAP	http://www.protomap.cs.huji.ac.il/
Megaclass	http://www.ibc.wustl.edu/megaclass/
• Protein domains	
ProDom	http://protein.toulouse.inra.fr/prodom.html
PRINTS	http://www.biochem.ucl.ac.uk/bsm/dbbrowser/PRINTS/ PRINTS.html
DOMO	http://www.infobiogen.fr/~gracy/domo/
PFAM	http://genome.wustl.edu/Pfam/
BLOCKS	http://blocks.fhcrc.org/
• RNA/DNA	
Ribosomal Database Project	http://www.cme.msu.edu/RDP/
The rRNA WWW server	http://rrna.uia.ac.be/
ACUTS[a]	http://pbil.univ-lyon1.fr/acuts/ACUTS.html

[a] Ancient Conserved UnTranslated Sequences.

the HOVERGEN database compiles multiple alignments and phylogenetic trees for all families of vertebrate protein-coding genes along with the corresponding GenBank annotations (79). This database provides all the data necessary to decipher the orthology/paralogy relationships among vertebrate multigenic families and is thus particularly useful for phylogenetic studies or for comparative analysis of vertebrate genes. However, this approach is limited to relatively well-conserved sequences alignable over their entire length. Conversely, databases of protein domains may achieve to cluster very distantly related sequences and are useful to analyse the structure, function, and evolution of modular proteins. For some complex families, it may be useful to consult specialized databases such as those available for immunoglobulins or HOX proteins (for a complete list, see the WWW page maintained by Amos Bairoch: http://www.expasy.ch/alinks.html).

7 Summary

In this chapter, we describe methods commonly used to align homologous sequences. Searching for the best alignment consists of finding the one that represents the most likely evolutionary scenario (substitutions, insertion, and deletion). Different alignment algorithms have been developed, but none of them is ideal. Because of time and memory requirements, algorithms that guarantee to find the best alignment for a given evolutionary model can be used in practice only with a very limited number of short sequences. Therefore, non-optimal algorithms based on heuristics have been proposed to gain speed and limit memory requirements. We discuss the choice between these different methods (progressive global alignment, block-based global alignment, motif-based local multiple alignment) according to the nature of the sequences to align. We also describe graphical tools that have been developed to visualize and edit multiple alignments. Finally, we mention several databases that compile multiple alignments of homologous protein or nucleotide sequences. All internet addresses where the tools and resources described here are available are listed in the following WWW page:

http://pbil.univ-lyon1.fr/alignment.html

References

1. Kimura, M. (1983). *The neutral theory of molecular evolution*, Cambridge University Press, Cambridge.
2. Doolittle, R. F. (1994). *Trends Biochem. Sci.*, **19**, 15.
3. Wootton, J. C. and Federhen, S. (1993). *Computers Chem.*, **17**, 149.
4. Altschul, S. F. and Gish, W. (1996). In *Methods in enzymology* (ed. R. F. Doolittle), Vol. 266, p. 460. Academic Press, London.
5. Patthy, L. (1996). *Matrix Biol.*, **15**, 301.
6. Schwartz, R. M. and Dayhoff, M. O. (1978). In *Atlas of protein sequence and structure* (ed. M. O. Dayhoff), p. 353. Nat. Biomed. Res. Found., Washington DC.
7. Gonnet, G. H., Cohen, M. A., and Benner, S. A. (1992). *Science*, **256**, 1443.
8. Henikoff, S. and Henikoff, J. G. (1992). *Proc. Natl. Acad. Sci. USA*, **89**, 10915.
9. Thompson, J. D., Higgins, D. G., and Gibson, T. J. (1994). *Nucleic Acids Res.*, **22**, 4673.
10. Overington, J., Donnelly, D., Johnson, M. S., Sali, A., and Blundell, T. L. (1992). *Protein Sci.*, **1**, 216.
11. Bains, W. (1992). *Mutat. Res.*, **267**, 43.
12. Hess, S. T., Blake, J. D., and Blake, R. D. (1994). *J. Mol. Biol.*, **236**, 1022.
13. Benner, S. A., Cohen, M. A., and Gonnet, G. H. (1993). *J. Mol. Biol.*, **229**, 1065.
14. Gu, X. and Li, W. H. (1995). *J. Mol. Evol.*, **40**, 464.
15. Benson, D. A., Boguski, M. S., Lipman, D. J., Ostell, J., Ouellette, B. F. F., Rapp, B. A., *et al.* (1999). *Nucleic Acids Res.*, **27**, 12.
16. Stoesser, G., Tuli, M. A., Lopez, R., and Sterk, P. (1999). *Nucleic Acids Res.*, **27**, 18.
17. Bairoch, A. and Apweiler, R. (1999). *Nucleic Acids Res.*, **27**, 49.
18. Barker, W. C., Garavelli, J. S., McGarvey, P. B., Marzec, C. R., Orcutt, B. C., Srinivasarao, G. Y., *et al.* (1999). *Nucleic Acids Res.*, **27**, 39.
19. Schuler, G. D., Epstein, J. A., Ohkawa, H., and Kans, J. A. (1996). In *Methods in enzymology* (ed. R. F. Doolittle), Vol. 266, p. 141. Academic Press, London.
20. Etzold, T. and Argos, P. (1993). *CABIOS*, **9**, 49.
21. Gouy, M., Gautier, C., Attimonelli, M., Lanave, C., and Di-Paola, G. (1985). *Comp. Appl. Biosci.*, **1**, 167.
22. Pearson, W. R. and Lipman, D. J. (1988). *Proc. Natl. Acad. Sci. USA*, **85**, 2444.
23. Altschul, S. F., Gish, W., Miller, W., Myers, E. W., and Lipman, D. J. (1990). *J. Mol. Biol.*, **215**, 403.
24. Altschul, S. F., Madden, T. L., Schaffer, A. A., Zhang, J. H., Zhang, Z., Miller, W., *et al.* (1997). *Nucleic Acids Res.*, **25**, 3389.
25. Altschul, S. F., Boguski, M. S., Gish, W., and Wootton, J. C. (1994). *Nature Genet.*, **6**, 119.
26. Hughey, R. and Krogh, A. (1996). *Comput. Appl. Biosci.*, **12**, 95.
27. Notredame, C. and Higgins, D. G. (1996). *Nucleic Acids Res.*, **25**, 4570.

28. Chan, S. C., Wong, A. K. C., and Chiu, D. K. Y. (1992). *Bull. Math. Biol.*, **54**, 563.

29. Carillo, H. and Lipman D. (1988). *SIAM J. Appl. Math.*, **48**, 1073.

30. Altschul, S. F., Carroll, R. J., and Lipman, D. J. (1989). *J. Mol. Biol.*, **207**, 647.

31. Gotoh, O. (1995). *Comput. Appl. Biosci.*, **11**, 543.

32. Sankoff, D. (1975). *SIAM J. Appl. Math.*, **78**, 35.

33. Needleman, S. B. and Wunsh C. D. (1970). *J. Mol. Biol.*, **48**, 443.

34. Lipman, D. J., Altschul, S. F., and Kececioglu, J. D. (1989). *Proc. Natl. Acad.. Sci. USA*, **86**, 4412.

35. Gupta, S., Kececioglu, J. D., and Schäffer, A. (1995). *J. Comput. Biol.*, **2**, 459.

36. Jiang, T., Lawler, E. L., and Wang, L. (1994). *ACM Sympos. Theory Comput.*, **26**, 760.

37. Feng, D. F. and Doolittle, R. F. (1987). *J. Mol. Evol.*, **25**, 351.

38. Taylor, W. R. (1988). *J. Mol. Evol.*, **28**, 161.

39. Higgins, D. G. and Sharp, P. M. (1988). *Gene*, **73**, 237.

40. Sneath, H. A. and Sokal, R. R. (1973). *Numerical taxonomy*. W. H. Freeman. San Francisco.

41. Saitou, N. and Nei, M. (1987). *Mol. Biol. Evol.*, **4**, 406.

42. Gotoh, O. (1996). *J. Mol. Biol.*, **264**, 823.

43. Vingron, M. and Argos, P. (1989). *Comput. Appl. Biosci.*, **5**, 115.

44. Sobel, E. and Martinez, H. (1986). *Nucleic Acids Res.*, **14**, 363.

45. McCreight, E. M. (1976). *J. ACM*, **232**, 262.

46. Vingron, M. and Argos, P. (1991). *J. Mol. Biol.*, **218**, 33.

47. Morgenstern, B., Dress, A., and Werner T. (1996). *Proc. Natl. Acad. Sci. USA*, **93**, 12098.

48. Morgenstern, B., Atchley, W. R., Hahn, K., and Dress, A. (1998). *Ismb*, **6**, 115.

49. Zhang, Z., Raghavachari, B., Hardison, R., and Miller, W. (1996). *J. Comput. Biol.*, **1**, 217.

50. Myers, E. and Miller, W. (1995). In *Proc. 6th ACM-SIAM Symposium On Discrete Algorithms*, p. 38.

51. Zhang, Z., He, B., and Miller, W. (1996). *J. Disc. Appl. Math.*, **71**, 337.

52. Abdeddaïm, S. (1997). In *Lecture notes in computer science*, Vol. 1264, p. 167. Springer–Verlag.

53. Smith, R. F. and Waterman, M. S. (1981). *J. Mol. Biol.*, **147**, 195.

54. Waterman, M. S. and Jones, R. (1990). In *Methods in enzymology* (ed. R. F. Doolittle), Vol. 183, p. 221. Academic Press, London.

55. Schuler, G. D., Altschul, S. F., and Lipman, D. J. (1991). *Proteins*, **9**, 180.

56. Lawrence, C. E., Altschul, S. F., Boguski, M. S., Liu, J. S., Neuwald, A. F., and Wootton, J. C. (1993). *Science*, **262**, 208.

57. Henikoff, S., Henikoff, J. G., Alford, W. J., and Pietrovoski, S. (1995). *Gene-COMBIS, Gene*, **163**, 17.

58. Bailey, T. L. and Elkan, C. (1995). *Ismb*, **3**, 21.

59. Lawrence, C. E. and Reilly, A. A. (1990). *Proteins*, **7**, 41.

60. Cardon, L. R. and Stormo, G. D. (1992). *J. Mol. Biol.*, **223**, 159.

61. McClure, M. A., Vasi, T. K., and Fitch, W. M. (1994). *Mol. Biol. Evol.*, **11**, 571.

62. Briffeuil, P., Baudoux, G., Lambert, C., De Bolle, X., Vinals, C., Feytmans, E., *et al.* (1998). *Bioinformatics*, **14**, 357.

63. Morrison, D. A. and Ellis, J. T. (1997). *Mol. Biol. Evol.*, **14**, 428.

64. Huang, X. and Miller, W. (1991). *Adv. Appl. Math.*, **12**, 337.

65. Thompson, J. D., Gibson, T. J., Plewniak, F., Jeanmougin, F., and Higgins, D. G. (1997). *Nucleic Acids Res.*, **25**, 4876.

66. Jeanmougin, F., Thompson, J. D., Gouy, M., Higgins, D. G., and Gibson, T. J. (1998). *Trends Biochem. Sci.*, **23**, 403.

67. Brocchieri, L. and Karlin, S. (1998). *J. Mol. Biol.*, **276**, 249.

68. Galtier, N., Gouy, M., and Gautier, C. (1996). *Comput. Appl. Biosci.*, **12**, 543.

69. Parry-Smith, D. J., Payne, A. W., Michie, A. D., and Attwood, T. K. (1998). *Gene*, **221**, GC57.

70. Schneider, T. D. and Stephens, R. M. (1990). *Nucleic Acids Res.*, **18**, 6097.

71. Srinivasarao, G. Y., Yeh, L. S. L., Marzec, C. R., Orcutt, B. C., Barker, W. C., and Pfeiffer, F. (1999). *Nucleic Acids Res.*, **27**, 284.

72. Yona, G., Linial, N., Tishby, N., and Linial, M. (1998). *Ismb*, **6**, 212.

73. Corpet, F., Gouzy, J., and Kahn, D. (1999). *Nucleic Acids Res.*, **27**, 263.

74. Attwood, T. K., Flower, D. R., Lewis, A. P., Mabey, J. E., Morgan, S. R., Scordis, P., *et al.* (1999). *Nucleic Acids Res.*, **27**, 220.

75. Gracy, J. and Argos, P. (1998). *Bioinformatics*, **14**, 164.

76. Gracy, J. and Argos, P. (1998). *Bioinformatics*, **14**, 174.

77. Bateman, A., Birney, E., Durbin, R., Eddy, S. R., Finn, R. D., and Sonnhammer, E. L. L. (1999). *Nucleic Acids Res.*, **27**, 260.

78. Henikoff, J. G., Henikoff, S., and Pietrokovski, S. (1999). *Nucleic Acids Res.*, **27**, 226.

79. Duret, L., Mouchiroud, D., and Gouy, M. (1994). *Nucleic Acids Res.*, **22**, 2360.

80. Maidak, B. L., Cole, J. R., Parker Jr, C. T., Garrity, G. M., Larsen, N., Li, B., *et al.* (1999). *Nucleic Acids Res.*, **27**, 171.

81. Van de Peer, Y., Robbrecht, E., de Hoog, S., Caers, A., de Rijk, P., and de Wachter, R. (1999). *Nucleic Acids Res.*, **27**, 179.

82. De Rijk, P., Robbrecht, E., de Hoog, S., Caers, A., Van de Peer, Y., and de Wachter, R. (1999). *Nucleic Acids Res.*, **27**, 174.

83. Duret, L., Gasteiger, E., and Perrière, G. (1996). *Comput. Appl. Biosci.*, **12**, 507.

Chapter 4

Hidden Markov models for database similarity searches

Ewan Birney

The Sanger Centre, Wellcome Trust Genome Campus, Cambridge, UK.

1 Introduction

Despite the huge number of genes in an organism, the protein coding genes are thought to be made from a limited number of basic protein structures. Evolution has reused these protein structures, combining them to form different proteins, and altering them in different genes to achieve different functions. The diversity of species, each with its own copies of genes made from the limited number of building blocks, means that, for a protein of interest, a number of different related proteins may be found. In this chapter, I will discuss one set of techniques which can be used to take advantage of this diversity of protein sequence. These techniques are all related to the use of *profiles*, which are also discussed in Chapter 5. In this chapter, the emphasis will be on the use of *hidden Markov models* (HMMs) for profile analysis. Some practitioners consider profiles to be a type of HMM.

It is important to realize that a protein might be related to another protein in a variety of different ways. It could be that the entire protein is homologous (that is, derived from a common ancestor) to another, such as human and mouse *src* protein (see *Figure 1*) or the human *src2* protein which is a paralog to the *src* protein. Alternatively only a portion of the protein might be derived from a common ancestor, such as the *fyn* protein, which shares a common C terminal region with a divergent N terminus to the *src* protein. Finally only a small region might be conserved, such as the SH3 domain which is also found in the Grb2 protein (along with many other proteins) with no other organization conserved between the two proteins. This last type of conservation, conservation of a *domain* generally corresponds to a structural domain of the protein which can fold independently and, in most cases, function independently of other regions. Figuring out when you have really defined a domain rather than a more extensive piece of conservation is one of the challenges for a researcher. Profile analysis is useful for all these different types of conservation. It is especially useful for domain analysis as this is the hardest feature to define using other methods.

(1)

(2)

(3)

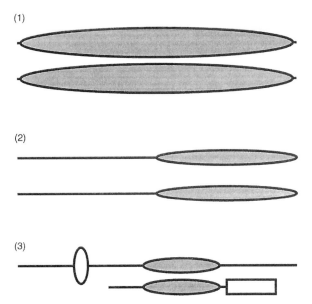

Figure 1 Three different types of relationship are shown. The grey ovals indicate regions which are conserved, whereas the lines and other boxes show regions which are not related. (1) Two very closely related genes, where the entire protein sequence in each gene is conserved. (2) Two genes where the C termini are related but the N termini are unique. (3) Two genes which share one domain but the other regions are entirely different.

2 Overview

For people coming from outside the field, the use of profiles and profile-HMMs can require confronting much confusing jargon and cryptic computer programs. This chapter is meant to demystify this type of analysis. The first point to emphasize is that the programs are, basically, just employing some concept of a 'consensus'. This follows intuitively from the observation that if some sequences have an Aspartate before a critical catalytic residue and others a Glutamate then a new enzyme can be expected to have either an Aspartate or a Glutamate at this position. This sort of simplistic rule is recast into a mathematically convenient form: resulting in some idea of a probability for each possible amino acid at a different position, called a *profile*. The difficulty lies, as in many areas in sequence analysis, that there may be different numbers of amino acids between conserved residues. A consequence of the differing lengths is that there are usually a number of different ways of providing a match to a 'consensus', and some way of choosing the 'best' one must be decided. The variable lengths between conserved residues also makes the statistical behaviour of the technique very hard to handle using conventional statistical analysis.

This chapter will concentrate first on using databases of profile-HMMs through the World Wide Web (WWW), which is by far the easiest way of using them. Then we will concentrate on PSI BLAST (1) which is the easiest do-it-yourself profile method, also available through the Web. The final example will

Table 1 Some useful Web addresses

Protein family sites	
Pfam (Europe)	http://www.sanger.ac.uk/Software/Pfam
Pfam (USA)	http://pfam.wustl.edu/
Prosite Profiles	http://www.isrec.isb-sib.ch/profile/profile.html
Prints	http://www.biochem.ucl.ac.uk/dbbrowser/PRINTS
BLOCKS	http://www.blocks.fhcrc.org/
SMART	http://coot.embl-heidelberg.de/SMART/
Software tools	
PSI-BLAST	http://www.ncbi.nlm.nih.gov/cgi-bin/BLAST/nph-psi_blast
HMMER2	http://hmmer.wustl.edu/

cover the use of the HMMER2 (2) package which I find to be the most effective profile-HMM package available. It is UNIX based and relatively easy to use, though it is not currently available through the Web. Finally I will outline some of the theories behind profile HMMs from the point of view of how it impacts on their practical use.

The reader should be aware that there are many other profile-HMM packages. I would draw your attention in particular to the Meta-MEME package (3) and the PROBE package (4) as well thought out solutions. There are also a number of other profile packages (5, 6) which are more focused on the use of the package by their own groups. Finally, a number of commercial solutions exist (7–9), and you may well have access to them. If you know someone on site who is already skilled in using one of these packages, it is best to use that local knowledge and treat this chapter as more of an introduction to the concepts involved. In addition, it is likely that when you are reading this chapter, that new methods or new presentations of old methods will become available. To keep up to date, use the web, and try the URLs in *Table 1* to find the most up-to-date resources.

Finally, I would like to warn users that I have a strong bias towards using a probabilistic framework to explain and justify the methods: this fits easiest with the HMM formalism and the use of Bayesian statistics (a branch of probability analysis). Other researchers are less zealous about using this sort of framework to explain the results. In either case the most important question is whether these methods are biologically useful, whatever the theories say.

3 Using profile and profile-HMM databases

The starting-point of this sort of analysis is usually a protein sequence which you might have derived from your own sequencing project of a gene of interest. The aim is to use a pre-made profile-HMM, previously constructed by another group to highlight regions of your protein which have homology to already well characterized domains. All these resources focus on domains, being the conserved building blocks of proteins, so the end result of the analysis will be to return which regions of your protein look as if they have particular domains.

Once you have a protein sequence it is probably best to put it into Fasta format (see *Section 8.5*), though many resources will allow you to use other formats. Then connect to one of the resources shown in *Table 1* and find the page for searching with your sequence.

Choose the 'search' page. Then use the 'file-upload' button on the forms to submit your own sequence, and click 'submit' or 'run analysis'. The search against the database will probably take a little over a minute, and should not take more than 10 minutes. Each resource returns its own particular format of results, but what is generally reported is the type of the domain, the position in your protein as a start- and end-point, and some indication of how confident your can be of the hit. They all provide a nice graphical representation of the domain on your sequence as a cartoon of the sequence with different coloured or shaped regions indicating the different domains. Clicking on the graphic will take you usually to an in-depth description of the domain, which in many cases will contain links to other resources and literature references. How to interpret the precise results varies from resource to resource.

3.1 Pfam

Pfam (10) is a database of protein families and corresponding profile-HMMs. Pfam uses the HMMER2 package to provide tools for making the HMMs in the first place and then for searching them. A search against Pfam will provide you with three ways of deciding confidence in the matches. The first is a classical e-value (expectation value) which generally is considered significant if it is below 1.0 for individual searches. The second is the Bits score which is derived from the underlying scoring scheme used to score the match between the sequence and the profile-HMMs. It is related to the Bayesian inference of the probability of the match (for a deeper explanation of the statistics read *Section 8*). A final check is provided by a manually derived cut-off which an 'expert' has chosen to separate the true examples from false examples. These cut-offs are chosen conservatively so that, to the researcher's knowledge, they do not misclassify any protein. This can mean that, in some cases, known trues are missed using this cut-off.

At the time of writing the Pfam database (Version 3.3) had 1344 protein families, which covered 57% of the protein primary sequence database. In new genome projects over one third of proteins had at least one hit to a protein family.

3.2 Prosite profiles

Prosite profiles (11) are an addition to the Prosite resource to define protein domains using profile-HMM technology. Prosite profile reports a classical e-value type statistic which is presented on a log scale. Scores above 5 can be considered significant and scores above 7 very significant. The raw score is not a meaningful number except for different examples of the same domain the better the score the better the match. There are no manually set cut-offs.

At the time of writing, there were 205 prosite profiles. There is no reliable way of estimating the coverage of prosite profiles. However, a researcher can combine a prosite profile search with a Pfam search, allowing the two resources to be combined in the same submission, with a common output. You are advised to make sure the prosite form is using the most up-to-date Pfam release.

3.3 SMART

SMART (12) is currently based on conventional, non HMM profile technology. The raw score is meaningless, rather you must trust the manually set cut-offs provided internally. At the time of writing there were 302 profiles in SMART. SMART is not focused on coverage but rather on providing very accurate alignments and resources of the domains of interest. It is likely by the time of reading this that SMART has switched to using the HMMER2 package rather than old style profiles.

3.4 Other resources and future directions

There are a number of other resources which provide access to ready made homology databases; in particular PRINTS (13) and BLOCKS (14) (see *Table 1* and Chapter 5). Both these resources are less focused on finding individual domains and instead focus on finding smaller 'motifs'. They may give less clear-cut answers, and for difficult domains may be less sensitive, but come with a number of useful utilities and options.

By the time of reading this chapter, it is likely that a number of resources will be using a common documentation resource (Interpro). This will provide more consistent documentation between the different resources, and is likely to be a forerunner to further integration between the resources.

3.5 Limitations of profile-HMM databases

The obvious limitation of a profile HMM database is that if the domains in your protein are not represented in the database then the databases will (hopefully) return nothing. The only option here is to start your own profile analysis using one of the techniques listed below. In addition, it might be that there is an error in the database, giving the wrong start/end points or misclassifying a region. In all the above databases, the most likely error will be that you miss a true domain in your protein. You can lower the thresholds for determining whether a domain exists or not, but be careful that you do not simply accept a false match due to 'noise'. Use the e-value statistic to decide whether this domain is justified on statistical grounds and read the information in section 7.0 on validating matches.

4 Using PSI-BLAST

PSI-BLAST (1) is a profile building and searching package which is fast, accessible through the Web, and aimed at a less expert audience than the other profile packages. This makes it ideal for occasional use or quick investigations of a particular protein sequence.

In many ways PSI-BLAST follows the same methodology as using the HMMER package below, just that this is done behind the scenes. PSI-BLAST starts from a single sequence, which is then searched against a database using the fast BLAST method. The resulting matches are aligned back to the query sequence, and this derived multiple alignment is used to estimate a profile. The profile is then used to search the database, collect homologues, and align back to the profile, and so the process iterates onwards until it stabilizes or some cut-off is exceeded.

A URL to start the process off is given in *Table 1*. You load in your protein sequence and launch the first search. At the end of each search you have the option of including or rejecting each sequence for the next iteration. This gives you the chance to eliminate potential false positives and include weak but true matches from your knowledge of the biology. An e-value statistic is provided to give an automatic selection of the next round of sequences, which should guide you in your selection.

Many of the problems inherent in using PSI-BLAST are also present when using HMMER, and so I would encourage you to read sections 6.0 and 7.0 carefully. Crucially you must be aware that the statistic to quote for the significance of a match is the first one in which it appears in the profile: once a particular sequence has been included in the set which makes the profile, it will, unsurprisingly, score very well against the resulting profile.

The other problems of PSI-BLAST are less to do with the method and more to how it is used. Because it starts with a single sequence, it is tempting to put in an entire sequence of interest and simply start iterating. If the sequence contains one common domain, although PSI-BLAST will find all the homologues of the sequence, both including the domain and excluding it your results will be dominated by this domain and become unmanageable. As you focus your effort on a particular region, it is better to excise that region and use that as a starting point for further analysis.

5 Using HMMER2

HMMER2 (2) is a package of UNIX command line programs which make and use profile HMMs. If you have no experience of the UNIX command line, then using HMMER2 is going to be a struggle. I suggest taking a short course in UNIX first. In addition to the HMMER2 software, you will need a number of other reasonably standard bioinformatics resources. In particular:

(a) A copy of an up-to-date protein database in fasta format as a single file.

(b) A method of retrieving sequences from this database, preferably with the ability to retrieve only a portion.

(c) A multiple alignment program such as Clustal W (17).

(d) A specialized multiple alignment editor.

It may also be useful to have some experience of a text reformatting language such as Perl or Python, or access to someone who can write small glue programs for you. Installing the resources is best done with the co-operation of the systems support group for the UNIX machine you are using.

5.1 Overview of using HMMER

Figure 2 gives the basic flow of profile analysis. The main steps are to create a multiple alignment of the region of interest and from this multiple alignment make a profile-HMM. The profile-HMM is searched against a protein database: a number of new protein matches may be identified and these can then be incorporated into the multiple alignment. There are two places where human knowledge can make a large difference in the analysis; firstly manual editing of the multiple alignment can produce dramatically better results, secondly which sequences are included or not into new alignments can be vetted using biological knowledge of the process.

5.2 Making the first alignment

To start the whole procedure off one needs to both identify potential homologues and produce the first multiple alignment. The first potential homologues are usually found using single sequence searches. Then these homologues are aligned using a multiple alignment program such as clustal w. This is discussed in depth in Chapter 3. An important issue to realize is that in many cases you will be attempting to make a multiple alignment of a *domain* common to a number of different proteins. To successfully make such a multiple alignment you will need to determine the rough boundaries of the domain in each protein from the single sequence searches. You will then need to excise the regions of the protein with a small (10 residue or so) leeway on each side: hopefully your database retrieval program will have this ability built in. Once the multiple alignment has been made you will probably want to edit it.

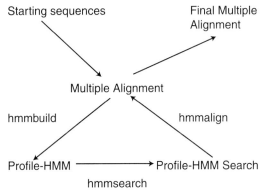

Figure 2 A flow diagram of how profile-HMMs are commonly used. The programs in the HMMer package which are used to provide the different transformations are given beside the arrows. PSI-Blast uses the same principles although much of the mechanics are then hidden from the user

5.3 Making a profile-HMM from an alignment

Thankfully the latest version of HMMER has sensible defaults for making profileHMMs. The program hmmbuild is used to make the HMM, which is run with defaults by typing at the UNIX prompt: 'hmmbuild HMM Alnfile' where alnfile is the name of a file containing a multiple alignment. The defaults which are run are as follows:

(a) **Tree weighting**. A tree is calculated from the multiple alignment and sequences weighted by how many close neighbours they have. This means that overrepresentation by one sub family (for example, many haemoglobin alpha chains for a globin profile-HMM) does not violently bias the profileHMM towards that subfamily.

(b) **Dirchlet mixtures**. A concept of accepted amino acid conservation patterns, such as Valine, Isoleucine, and Leucine being common replacements for each other is provided as Dirchlet mixtures. For more information about Dirchlet mixtures, see *Section 8.4*.

(c) The effective number of sequences is estimated, meaning that alignments which are of very close homologues will be assumed to represent fewer sequences than alignments of more distantly related sequences (see *Section 8.4*).

(d) The placement of which columns in the multiple alignment are match columns as opposed to insert columns.

These defaults are typical for the needs of most HMMs. As you get used to the package, you may want to change them (options are provided on the command line; to get a full list of options type 'hmmbuild –help'), though even experts tend not to deviate too much from these settings.

The only parameter which is worth altering routinely is the local/global mode switch. By default the profile-HMMs are built in global mode, where the entire HMM must be matched against the sequence. An alternative is local mode, by using the -f flag on hmmbuild, which allows only a portion of the HMM to match. If the global mode is used, the HMM becomes far more sensitive for finding distant members, but is unsuitable for finding fragments. If you have misdefined the region of interest, perhaps making it too long at the N or C termini, then the global mode will penalize sequences that do not fit the entire model, preventing them being found. Generally global mode is a better choice, but at the start of the analysis, local mode can be a more sensible choice as one is not always very confident about the locations of the ends of a domain.

Having made the profile-HMM, you need to calibrate it by typing 'hmmcalibrate HMM'. The calibration step calculates statistical parameters for the HMM by generating a random database. This statistical parameterization is crucial for its effective use. Calibration takes around 10 minutes or so.

5.4 Finding homologues and extending the alignment

The HMM can be searched against a database of sequences using the program hmmsearch. This takes the HMM file as the first argument and the database file as the second argument. The database file is in Fasta format. The results are printed on standard output, so you usually need to redirect the output to save that information in a file. The results give you the following information:

(a) A list of sequences which the HMM hit, ranked from most significant to least.

(b) A list of domains contained in the sequences, ranked from most significant to least.

Notice that a particular sequence can contain more than one domain. In particular, although each of the domain scores might, on their own, not be significant, the combined score of multi-domain match might easily be so.

Both the per-sequence and per-domain matches are provided with two statistics: a bits score and an e-value (see *Section 8.2*). The more reliable score is the e-value; e-values down to 1.0 can be considered significant (an e-value of 1.0 means that, by chance, 1 random sequence is expected to get this score in the database of the size which you used). The bits score is helpful as it is independent of database size.

Having chosen a significance level one would then like to make a new multiple alignment of all the protein sequences found by the HMM. At the moment, this is the most labour intensive step, as the HMMER package does not provide all the functionality for this task. Somehow, one needs to extract all the sequences which are hit and truncate them to the correct start/end points. This is best done by a perl script or similar device. Once you have all the sequences which were hit, as a Fasta file, the program hmmalign will provide a multiple alignment of the proteins on the basis of the HMM. This multiple alignment can then be used to make a new HMM for the next round.

6 False positives

One of the problems inherent to the iterative procedures, both PSI-blast and the use of HMMER outlined above is that if a false positive is added to the alignment, itself and any close relatives will score highly against the profile-HMM. For example, if you inadvertently add a globin sequences to a protein kinase alignment the resulting HMM will match globin sequences surprisingly well.

This ability to start collecting false positives at will means that a researcher should ideally by very vigilant as the iterations progress. Indications that the profile might be picking up noise are:

(a) Low complexity regions occurring in alignment.

(b) A region overlapping a known domain, where it is clear that the multiple alignment is not a divergent subfamily for this domain.

(c) Biological information that indicates that this match is false.

7 Validating a profile-HMM match

Once a researcher has found a suggested domain, how can they validate this? The score of the sequences to the HMM from the final alignment is not the correct measure of the significance of the match, as it includes all the sequences you wish to score, and they will all score well. In fact the problem of justifying a grouping of sequences is not well handled by the current statistics, in particular when an iterative strategy is used. The following lines of evidence may be used to give a researcher confidence that the similarity they observe is not by chance.

(a) See whether all the sequences can be connected together by significant single sequence scores (e.g. from programs such as BLAST2). Ideally one should be able to show this with the full length proteins (just taking the domain improves the statistics considerably).

(b) Quote the significance of the 'new' sequences for the first time they provided a significant score against the profile-HMM.

(c) To show that A is related B, show that by starting from either A or B one can produce a profile which finds the other sequence using criterion (b).

(d) Provide biological justification that the relationship makes sense (e.g. common mode of enzymatic action). Conversely, biological information which indicates that they should not be related should lessen the researcher's belief in the result.

8 Practical issues of the theories behind profile-HMMs

8.1 Overview of profile-HMMs

A profile is a sequence of conserved positions, each conserved position having a score for each amino acid. For practical purposes, one should neither assume that a particular sequence has all the conserved positions in a profile, nor that a particular sequence will not introduce additional amino acids between two conserved positions in the profile. These two possibilities are the two types of gaps, that is a deletion of part of the profile, or an insertion of residues relative to the profile. A number of *ad hoc* methods have been produced to solve the gap problem in ways that seem biologically sensible.

The *ad hoc* nature of profiles was replaced by a stronger theory based around HMMs which still essentially produced the same sort of profile as before. An HMM is a mathematical model which produces a stream of some observable information in a probablistic manner. In the case of protein sequences, the observable information will be the amino acids. The probablistic nature of the process of producing the amino acids means that a particular HMM can produce more than one set of protein sequences, and also that different sequences are produced with different probabilities. The hidden part of a HMM is that one does not know which model made the particular amino acids one is looking at, and, if one did know the model, which amino acids were made by which part of the model. It is these two problems which one wants to solve, and they correspond to the two questions: 'does this sequence have an example of this HMM in it' and 'if this sequence does have an example of the HMM, what is the alignment of the HMM to the sequence'.

The HMMs which have been used in this field deliberately mimic the profile model described above. For each conserved column in the multiple alignment, three possible states are permitted: a match state which indicates that a single residue is being aligned to the position, a delete state, which indicates that no amino acid in this protein is present for this position, and an insert state, which allows any number of amino acids to be inserted after the match position. A full-length HMM will have some 500 or so different states, broken down into triplets representing conserved column positions. The behaviour of these states is governed by probabilities for the production of different amino acids from match and insert states and probabilities for the transitions to the neighbouring states. These probabilities are analogous to the scores of the profile and the gap penalities in the profile respectively. Indeed, for practical purposes, the probability representation of a profile-HMM is rarely used. Instead, the probabilities are transformed into sensibly sized integers via a log transformation. In this logged representation, adding the numbers is equivalent to multiplying the underlying probabilities, making the correspondence between profiles and profile-HMMs all the more clear.

Given a particular profile-HMM, the questions 'does this sequence have an example of my HMM' and, given that the last question is true, 'what is the alignment of the sequence to the HMM' can be easily answered using some well known algorithms. These two questions are essentially what hmmsearch and hmmalign provide answers for in the HMMER package.

8.2 Statistics for profile-HMM

Every sequence will match the profile HMM in some manner: some sequences will match the profile HMM better (in the sense that the probability that the HMM would produce such a sequence is higher) than others. How does this statistic (called a likelihood) allow you to say whether this match really is due to the presence of this domain or is it just by chance?

As the underlying basis for HMMs is a probabilistic model, this question is easily answered by Bayesian statistical methods. Non-mathematicians usually have not been exposed to Bayesian statistics previously. Bayesian statisitics try to answer such questions by assigning a probability to its being true or not. In contrast classical statistics answers the question by assuming a particular hypothesis, which is rejected or not by the data. Both approaches are guaranteed to give the same answer when enough data is taken into account, but will often provide different answers for practical problems.

To provide a Bayesian interpretation of the profile-HMM, all possible different models of how the sequence was produced, have to be defined: for profile-HMMs, two models will be considered, the model of the profile-HMM domain and a *null* model of random amino acids drawn from the frequencies found in a large protein database. The probability of seeing the observed sequence under the assumption of the two different models are given, each being a likelihood.

What is quoted is the log-likelihood ratio of the two models: when the base of the log is 2, this statistic (the log-likelihood ratio) is called a *Bits score*. A bits score of 0 is when the likelihood ratio is 1, and hence each model is equally likely to have produced the sequence. Depending on how the ratio is quoted, either more negative or more positive scores indicate that the desired model is more likely. In the HMMER package, the more positive the bits score the better the match to the profile-HMM.

The likelihood ratio does not provide quite enough information to allow an estimate of the probability of the profile-HMM occurring, given the sequence seen. An additional piece of information, being the probability of the profile-HMM occurring without seeing the sequence data needs to be defined. As this information has to be defined without seeing any sequence, it is called *prior* information. Mathematically this is the same idea as the prior information which will be introduced in the next section, but in practice it is used in a very different aspect. Sensible priors include $1/d$ where d is the size of the data base or one could use the probability of a random sequence having this domain in a genome, say 1/10 000. Finally, to be confident of the match, the probability of the profile-HMM occurring should be over 0.95. These two extra manipulations —the prior information and the need for a significant probability translate into a bits cut-off above which one considers matches to be significant. 25 bits translates to sensible choices of prior and significance, and so matches over 25 bits can be considered to be significant.

There is another statistic that can be used to estimate whether the match is significant or not. This is a classical (or frequentist) statistic and is one that most users will be more familiar with. To provide a frequentist statistic, one needs to assume that the match is random, derive the probability that a random match would produce the score, and reject the assumption if this seems very unlikely to have occurred. The problem with this sort of analysis was that it was clear that the distribution of scores of random sequences against a profile-HMM was not normally distributed, and so estimation of probability was very difficult. In recent years the field has produced theoretical and empirical evidence that the distribution is closely related to an Extreme Value Distribution (EVD) (15). PSI-Blast assumes that for a particular way of making a profile, all profiles have the same EVD parameters, regardless of content. PSI-Blast therefore tabulates this information for all possible profile construction mechanisms, and uses the tabulated parameters. HMMER uses a separate calibration step, where the profile-HMM is compared to a large random database, and an EVD is fitted to the resulting distribution. The parameters from this fitting are stored in the HMM so they can be reused for individual sequence searches.

The natural way of reporting the classical statistic is as an expectation value (e-value). This is the number of sequences expected to get this score by chance, and is simply dr where d is data base size and r is the probability that a random sequence will get this score. An e-value of 1.0 is therefore where you expect to start seeing random sequences: e-values less than this are significant.

Which statistic to use: bits score or e-value? It is clear that the e-value statistic is more robust and more sensitive in the HMMER2 package, and it is what I would recommend. However, the e-value has some less desirable properties, in particular, it changes as the database size changes, unlike the bits score—of course, if you decide the prior on the bits score should be $1/d$ (d is database size), the cut-off for significance of the bits score will change with database size. Quoting both in publications is very sensible.

8.3 Profile-HMM construction

The use of a particular profile-HMM is well understood. A harder problem to solve is 'what profile-HMM best represents my collection of known family members'. By analogy with other fields, such as speech recognition, this problem can be answered by expectation maximization, or similar methods, where an HMM is constructed that maximizes the probability of producing all the sequences known to belong to a certain family. In theory this can work from just the sequences, and no multiple alignment, effectively both aligning the sequences and making the profile-HMM at the same time. In practice, training an HMM from unaligned sequences does not work well, principally because of local minima problems. A better solution is to train the HMM from already aligned sequences, as in *Figure 2*. When already aligned sequences are presented to the training program, the only aspect which the program must estimate is which columns to consider as conserved positions, and which columns should be collapsed into an insert state of the preceding conserved position. Indeed, if you so wish, you can provide this information directly. Once this is known, it is relatively easy to estimate the probabilities for the HMM from the observed sequences using standard theories.

8.4 Priors and evolutionary information

The final problem in HMM construction is that, in general, one is not interested in finding proteins which are only slightly different from the examples one already knows. Rather one wants to find a new subfamily related to the subfamily one has already gathered. This usually means that a sequence from the new subfamily will have some features which are not present in any of the sequences one has already gathered, and yet because of the pattern of conservation and knowledge of behaviour of proteins (for example, a conserved valine position is more likely to have a leucine in a distant subfamily member at this position, than an arginine) one can recognize it as being a related member.

The introduction of extra knowledge into the process of estimating an HMM is called *prior* information, indicating that it is known before any sequence data is seen. Ideally one would like to represent all the knowledge about protein evolution and protein structure in some manner which would allow the profile-HMM construction machinery to use it. In practice a number of assumptions have to be made to allow the mathematics of profile construction to work.

1. Profile-HMM building does not currently work with a concept of the observed sequences being related on an evolutionary tree. Therefore, if you present a

profile-HMM construction method with 10 near-identical alpha globins and one beta globin, the resulting HMM would be predominantly alpha globin. This is solved by weighting the sequences by a tree before applying the profile- building machinery. In the above example, each alpha globin sequence might get a weight of 1/10, and the beta globin sequence a weight of 1.

2. The estimation of amino acid probabilities, taking into account protein evolution, has a stronger theoretical backing. The problem is phrased as an under sampling problem, where although one has a column of, say, 10 amino acids at this position the frequencies of amino acids represented by observation is not an ideal way to estimate the underlying probabilities; clearly not all the amino acids can be represented even once! This problem occurs in many other situations and has been well studied. A good solution is to provide the estimation machinery with prior knowledge of what sort of amino-acid frequencies one expects in columns, for example, one with high leucine, valine, and isoleucine probabilities, and another with a high probability of arginine and lysine, but low probability for hydrophobic amino acids. These distributions are represented in a complicated mathematical form called Dirchlet mixtures (16) and, by using them, the estimation of probabilities for amino acid positions can take into account evolutionary information. A Dirchlet mixture is a just a convenient mathematical form for this information; there is nothing special about them except that they make the downstream mathematics far easier to handle. The Dirchlet mixture can be thought of rather like a protein comparison matrix used in ad hoc profile methods.

3. The final twist to profile-HMM construction is how to balance the information from the Dirchlet mixtures (the prior) with the information from the observed multiple alignment. The problem here is that the aim is to find new sub families so, the fact that one has seen over 1000 different alpha globins, does not mean that the observed amino acid frequencies on their own will make a good HMM for finding beta globin sequences. This is solved in the HMMER package by estimating how many effective sequences one has observed (a collection of divergent sequences will count more than a collection of close relatives).

8.5 Technical issues

In Fasta format the first line starts with the greater-than sign (>) and is followed by the name of the first protein, which should be composed of only characters from the alphabet, or the underscore symbol (_) or numbers. Most programs (though not all) will allow any non space character in the name. Following this line the next lines are the protein sequence in one letter code. The sequence stops at either the end of the file or the next greater-than (>) sign (which marks the start of the next sequence name if there is more than one sequence in the file).

References

1. Altschul, S. F., Madden, T. L., Schaffer, A. A., Zhang, J., Zhang, Z., Miller, W., *et al.* (1997). *Nucleic Acids Res.*, **25**, 3389.

2. Eddy, S. R. (1998). *Bioinformatics*, **14**, 755.

3. Grundy, W. N., Bailey, T., Elkan, C. P., and Baker, M. E. (1997). *Comput. Appl. Biosci.*, **13**, 397.

4. Neuwald, A. F., Liu, J. S., Lipman, D. J., and Lawerence, C. E. (1997). *Nucleic Acids Res.*, **25**, 1665.

5. Krogh, A., Brown, M., Mian, I. S., Sjolander, K., and Haussler, D. (1994). *J. Mol. Biol.*, **235**, 1501.

6. Bucher, P., Karplus, K., Moeri, N., and Hoffmann, K. (1996). *Comp. Chem.*, **20**, 3.

7. http://www.gcg.com/

8. http://www.netid.com/

9. http://www.compugen.com/

10. Bateman, A., Birney, E., Durbin, R., Eddy, S. R., Finn, R. D., and Sonnhammer, E. L. (1999). *Nucleic Acids Res.*, **27**, 260.

11. Hofmann, K., Bucher, P., Falquet, L., and Bairoch, A. (1999). *Nucleic Acids Res.*, **27**, 215.

12. Ponting, C. P., Schultz, J., Milpetz, F., and Bork, P. (1999). *Nucleic Acids Res.*, **27**, 229.

13. Attwood, T. K., Flower, D. R., Lewis, A. P., Mabey, J. E., Morgan, S. R., Scordis, P., *et al.* (1999). *Nucleic Acids Res.*, **27**, 220.

14. Henikoff, J. G., Henikoff, S., and Pietrokovski, S. (1999). *Nucleic Acids Res.*, **27**, 226.

15. Altschul, S. F. and Gish, W. (1996). In *Methods in enzymology* (ed. R. F. Doolittle), Vol. 266, p. 460. Academic Press.

16. Brown, M., Hughey, R., Krogh, A., Mian, I. S., Sjolander, K., and Haussler, D. (1993) In *Proceedings of the First International Conference on Inteligent Systems for Molecular Biology* (ed. L. Hunter, D. Searls, and J. Shavlik), p. 47. AAAI Press, Menlo Park, CA, USA.

17. Thompson, J. D., Higgins, D. G., and Gibson, T. J. (1994). *Nucleic Acids Res.*, **22**, 4673.

Chapter 5

Protein family-based methods for homology detection and analysis

Steven Henikoff and Jorja Henikoff

Howard Hughes Medical Institute, Fred Hutchinson Cancer Research Center, Seattle, USA.

1 Introduction

1.1 Expanding protein families

Most methods for homology detection have traditionally relied upon pairwise comparisons of protein sequences, and in recent years, several improvements in pairwise methods have been introduced (see Chapter 8). But, with sequence data becoming available at an accelerating rate, there is an increasing opportunity to use multiple related sequences for improved homology detection. Even when functional information is lacking for known members of a protein family, these members can be aligned and the alignments used in searches. Protein multiple alignments have been shown to improve performance of secondary structure prediction methods by identifying constraints on positions (1, 2), and so it seems reasonable to expect that improvements will be likewise obtained by using multiple alignments for homology detection and analysis. In this chapter we review some of the numerous methods that are aimed at achievement of this goal.

1.2 Terms used to describe relationships among proteins

Any region of shared similarity between sequences may be referred to as a 'motif'. To call something a motif does not necessarily imply that shared similarity reflects shared ancestry. For example, the well-studied helix-turn-helix DNA binding motif is found in proteins belonging to apparently unrelated families with different origins, and this suggests convergence towards a common structure.

Confusion often arises from the use of the structural term 'domain' to describe regions of sequence similarity. A separately-folded domain may be obvious from looking at the structure of a protein, but, without seeing a structure, it may not be possible to decide from an alignment what is the limit of a domain.

Furthermore, domains need not be contiguous along a sequence, and it is common for proteins to fold starting with one domain, continue on to fold into another domain, then return to the original domain further along the sequence. For sequence analysis applications, a useful concept is that of 'module'. A module can be thought of as a sequence segment that may be found in different contexts in different proteins, the result of mobility during protein evolution. Modules may correspond to separately folded domains, such as the C_2H_2 zinc finger motif, and they may be repeated within a sequence. Unlike domains, modules are necessarily contiguous along a sequence. Nevertheless, readers should be aware that modules identified by sequence similarity are typically referred to as 'domains' without confirming structural evidence, and the term 'multi-domain' is commonly applied to any chimeric protein.

'Family' is a generic term used to describe proteins (or genes) with sufficiently high sequence similarity that common ancestry may be inferred. A multi-domain protein might have modules that belong to several different families. Confusion can arise from the use of terms such as 'superfamily' and 'subfamily', which are not precisely defined. For these terms to be useful, some sense of what is meant by a family is required. Thus, if we refer to the opsins, the beta-adrenergic receptors and the olfactory receptors as separate families, even though they are related to one another, then we would refer to the G-protein coupled receptor superfamily to describe them all. Sometimes, proteins fold similarly, even though no sequence similarity between them is detected. For instance, the TIM barrel fold has been found for dozens of separate superfamilies, and it is not certain as to whether they share common ancestry. Conversely, sequence similarity may be evident, even though common ancestry is doubtful, as in the case of coiled-coil regions of proteins. As a practical matter, the methods described in this chapter are most useful for families and modules, where alignment-based methods can provide profound functional insights.

1.3 Alternative approaches to inferring function from sequence alignment

Opposing views of sequence alignment problems have resulted in two different classes of comparison tools for sequence analysis of protein families. Motif-based tools consider aligned protein sequences to consist of nuggets of alignment information (blocks) separated by regions that have no certain alignment. To proponents of this view ('blockers'), the task is to first find these conserved nuggets. 'Gappers' agree that there are nuggets worth finding, but that these will be best found by determining where to place the gaps in each sequence such that the blocks correctly align. Both blockers and gappers agree that aligning conserved nuggets is worthwhile, but they use different methods for accomplishing this. Blockers favour motif-based methods that first find regions of conservation. Such block-based methods as the BLAST family of searching programs and the BLOSUM amino acid substitution matrices continue to be favoured for many comparative sequence analysis applications (Chapter 8). Gappers favour methods that decide upon gap placement (described in Chapter 3) and use gap-based tools, especially dynamic programming and hidden Markov models (described in Chapter 4), for database probing. As is so often the case, the truth lies somewhere between the extremes. So although we blockers prefer to reduce the protein alignment problem to finding a set of ungapped blocks to represent a protein family or module, we recognize that insertions and deletions occur occasionally within conserved regions, and this is challenging for block-based methods.

Alignment usefulness is the major driving force in developing methodology. Obtaining a correct alignment is more important for some applications than for others. The ability to find corresponding residues and local regions that have similar functions is of unquestionable value, and the better the conservation of a residue or local region in a sequence, the more likely it is that common function can be inferred. Regions of uncertain alignment, such as those that different alignment programs using various score parameters disagree on, have little if any value for drawing functional inferences. However, so much alignment information is present in conserved regions that it might make sense to align beyond what can be done with confidence in order that more nuggets are captured. We suspect that this accounts for the success of many gap-based approaches: gapped alignments may have a high degree of uncertainty, but the proportion that is aligned successfully is sufficient to identify extensive shared regions of sequence similarity in database searches, even to the point of discovering correct folds more successfully than structure-based threading (3).

Practical utility requires ready availability to the general public. Nowadays, this means access via the World Wide Web using a browser, and so nearly all methods highlighted here (*Table 1*) can be performed without any special software, hardware, or computational expertise. Some potentially powerful tools are too computationally intensive to be made available in this way. Additionally, some tools require a specialist's knowledge and are not sufficiently automated for the average biologist to use them wisely. We believe that such tools should be avoided if possible: sequence alignment is fraught with hazards, and erroneous conclusions drawn from naive use of powerful sequence analysis tools abound (4).

2 Displaying protein relationships

2.1 From pairwise to multiple-sequence alignments

Depictions of pairwise sequence alignments are not easily extended to multiple alignments. For displaying pairwise alignments, identities and conservative replacements are typically emphasized using symbols between the aligned sequences. However, adding just a third sequence below the first two leaves open the question of how to represent similarities between the third sequence and the first, and addition of more sequences becomes increasingly complex. Dot matrix representations of pairwise alignments present the same problem.

Table 1 URLs

1. Displaying alignments	
Boxshade	http://www.ch.embnet.org/software/BOX_form.html
Logos	http://blocks.fhcrc.org/about_logos.html
Trees	http://blocks.fhcrc.org/about_trees.html
2. Finding alignments	
BCM launcher	http://dot.imgen.bcm.tmc.edu:9331/multi-align/multi-align.html
MACAW	ftp:/ncbi.nlm.nih.gov/repository
3. Searching family databases	
Prosite	http://www.expasy.ch/prosite/
Blocks	http://blocks.fhcrc.org/blocks_search.html
Prints	http://www.bioinf.man.ac.uk/fingerPRINTScan/
	bin/attwood/SearchPrintsForm2.pl
	http://blocks.fhcrc.org/blocks_search.html
ProDom	http://www.toulouse.inra.fr/prodom/doc/blast_form.html
Pfam	http://pfam.wustl.edu/hmmsearch.shtml
	http://www.sanger.ac.uk/Pfam/search.shtml
Proclass	http://www-nbrf.georgetown.edu/gfserver/genefind.html
ProfileScan	http://www.isrec.isb-sib.ch/software/PFSCAN_form.html
Identify	http://dna.Stanford.EDU/identify/
Recognize	http://dna.stanford.edu/ematrix/
Prof_Pat	http://wwwmgs.bionet.nsc.ru/mgs/programs/prof_pat/
4. Searching with multiple alignments	
MAST	http://meme.sdsc.edu/meme/website/mast.html
COBBLER	http://blocks.fhcrc.org/
PSI-BLAST	http://www.ncbi.nlm.nih.gov/cgi-bin/blast/psiblast.cgi
LAMA	http://blocks.fhcrc.org/LAMA_search.html

Such displays not only become complex, but also they fail to represent shared similarities. Because of these limitations, multiple alignment representations that emphasize regions of high similarity have been introduced.

Traditional displays of multiple sequence alignments show aligned sequences one above the next, highlighting identical or similar residues in a column using boxes, shading, or colour. These displays can be complex, especially when representing protein families that consist of large numbers of sequences that group into distinct subfamilies. Position-based representations greatly simplify the display of multiple alignments, because relationships between successive amino acids in a sequence are not shown. Indeed, computer programs that utilize multiple sequence alignment information in searches likewise consider all positions in aligned sequences to be independent of one another, and so position-based representations depict approximately what a searching program examines.

2.2 Patterns

The simplest position-based representations of multiple alignments are patterns, which display only key conserved residues. The Prosite database (5) is a compilation of sequence families that provides one or more patterns representing each family. An example of a Prosite pattern is C-x(2,4)-C-x(3)-[LIVMFYWC]-x(8)-H-x(3,5)-H, which is read as: cysteine, followed by 2 to 4 amino acids of any type, followed by cysteine, followed by any 3 amino acids, followed by one of the following—leucine, isoleucine, valine . . . , and so on. Although Prosite patterns are manually derived from multiple sequence alignments, the process of determining patterns from alignments has been automated (6, 7). Pattern-based methods are described in detail in Chapter 7.

2.3 Logos

Sequence logos (8) are vivid graphical displays of multiple sequence alignments consisting of ordered stacks of letters representing amino acids at successive positions (*Figure 1*). The height of a letter in a stack increases with increasing frequency (or probability) of the amino acid, and the height of a stack of letters increases with increasing conservation of the aligned position. Stack heights are displayed in bit units. One bit is the answer to a yes-or-no question, where yes is as likely as no. About 4 bits are required to fully specify a residue at a given position, because the first question narrows the field from 20 residues to 10, the second to 5, etc. The most probable amino acid is at the top of the stack, making it more visible, and below it is the next most probable residue, and so on. Logo colours or shades are chosen to emphasize similar amino acid properties. Logos can be scaled such that the stack height is proportional to the observed frequency of a residue divided by the frequency with which the residue is expected to occur by chance (odds ratio).

2.4 Trees

The most serious drawback of position-based displays is that they show only alignment information in common among the sequences in a family, not the

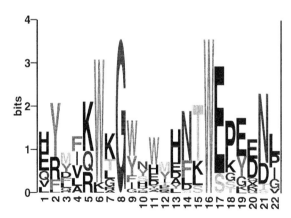

Figure 1 Sequence logo depicting the chromodomain block (BL00598 in Blocks v. 11.0). Each alignment position is represented as a stack of letters, where the height of the stack and the height of each letter is measured in bit units.

differences that distinguish between sequences or groups of sequences. In contrast, trees are designed to discriminate between individual sequences by providing an intuitive diagram of relationships drawn from an alignment. Although trees were introduced as phylogenetic tools, they have become increasingly popular for displaying protein families. Trees can generally be used to distinguish orthologs from paralogs, because orthologs will branch in a manner that is consistent with species phylogeny, whereas paralogs may deviate. Paralogous proteins often have distinct biological activities, and so trees can guide experimental investigations.

For using trees to draw inferences about function and not to infer phylogeny, some aspects of tree construction and interpretation that matter to phylogenetics may be relatively unimportant. Debate continues about how a tree should be constructed from an alignment, whether to use parsimony, distance, or maximum-likelihood methods. However, we are unaware of evidence that the choice of a tree-making program matters much for distinguishing paralogs from orthologs or for deciding whether one branch has a function that is comparable to the function of another branch. That is, we are not using a tree to distinguish whether bakers yeast is closer to fruit flies than to maize, so we can ignore the details at the leaves of a tree and focus on the separation of one group of yeast, fly, and maize proteins from paralogous groups of proteins. In our experience, the quality of the alignment might be important in making such distinctions, but different tree-making programs draw trees that are sufficiently similar for our purposes. Distance methods such as neighbour-joining (9) do have the advantage of being fast and can then be applied to large numbers of proteins and thus are suitable for making trees to analyse protein families.

3 Block-based methods for multiple-sequence alignment

Searching methods that utilize multiple sequence alignment information can be block-based or gap-based. Gap-based methods for finding multiple alignments are described in detail in Chapter 3, and gap-based methods for searching with them, such as hidden Markov models (HMMs), are described in Chapter 4. In this section, we describe block-based strategies for finding multiple-sequence alignments that are then used for database searching by many of the methods described in subsequent sections.

3.1 Pairwise alignment-initiated methods

One general approach to finding motifs involves performing pairwise comparisons between sequences and then asking which high-scoring local regions are in common for most or all of the sequences in the group. Where aligned segment pairs overlap, they are multiply aligned. However, determining which segments truly overlap can be challenging, and different methods have been introduced (10–12). In the MACAW program, overlapping segment pairs that exceed a threshold score are combined into an ungapped block (12). The extent of the block is limited by the requirement that each column have some minimum degree of homogeneity. Blocks that are separated by the same number of residues in all sequences may be fused, and so blocks can contain both conserved and diverged positions. MACAW is an interactive program that allows users to choose a set of blocks from among candidates. The threshold score for block searching can be relaxed by the user in order to find new blocks in regions between blocks that were found in the first pass.

Starting with pairwise alignments presents the same potential drawback as for gap-based hierarchical multiple sequence alignment programs (Chapter 5), which is that information in common for all of the sequences might not be represented in the pairwise alignments. In addition, the number of pairwise comparisons needed is n^2 for n sequences, and this can become somewhat impractical for large protein families and long sequences. Simultaneous methods for finding motifs, described below, can potentially avoid these problems.

3.2 Pattern-initiated methods

The rapidity with which amino acid 'words' can be scanned exhaustively through a set of related sequences has motivated pattern-based motif finders (13–17). An example of this approach, Motif (15), examines all sequences for the presence of spaced triplets of the form aa_1 d_1 aa_2 d_2 aa_3 where d_1 and d_2 are fixed distances between the amino acids. So Ala–Ala–Ala is one triplet, Ala–x–Ala–Ala is another, and Leu–x[16]–Ala–x[7]–Val is another. An exhaustive search is carried out for all such triplets in the full set of related sequences using all combinations of d_1 and d_2 out to a reasonable maximum distance (about 20). The rationale is that true motifs will typically include one or more sets of spaced triplets in all of the sequences in the group. Because some true motifs do not contain aa_1, aa_2, and aa_3 in the full set of sequences, the number of sequences required to contain a triplet (the 'significance level') can be reduced. In such cases, the block containing the triplet is scanned along each of the sequences that lack the triplet to find the best segment based on maximizing an overall score for the block. Each sequence is then rescanned to maximize the score. ASSET generalizes the search for patterns by scanning sequences for shared flexible patterns that occur in multiple sequences at a statistically significant level (17).

3.3 Iterative methods

Other approaches avoid limiting the motif search to a predetermined list without becoming computationally explosive by detecting motif 'seeds' that occur in as few as two sequences, then asking whether any of these seeds can mature to include other sequences in the group (18–21). Both Expectation-maximization (EM) and 'Gibbs sampling' (20) start with a block of specified width, then align random positions within all but one sequence. In EM, this sequence is scanned along the block, and the segment that maximizes a block score is chosen. In Gibbs sampling, the segment is chosen by a random sampling procedure, where the probability of being chosen is proportional to the block score. Other sequences are then sampled in the same way to further improve the significance of the alignment. Successive rounds of EM or Gibbs sampling continue until no further improvement is seen.

3.4 Implementations

Some of these methods are conveniently available over the internet, and sequences in FASTA format may be submitted by either pasting into a window or by file browsing. Because many real motifs can be subtle and as short as a few residues, sensitive methods may return alignments for sequences that are not based upon true relationships. A simple experiment (*Protocol 1*) demonstrates that even sequences chosen at random from a database can be aligned to yield motifs that appear convincing and will easily detect the parent sequences and their homologues from sequence databanks. Furthermore, even gross misalignments can be masked by the existence of significant similarity among just a fraction of sequences, and visual examination is notoriously unreliable (*Figure 2*). One solution is to report a reliability measure for each position (22), and one measure is implemented in Match-Box (23). BlockMaker's solution (24) is to apply two very different motif finders with different scoring systems, Motif and Gibbs sampling. In each case, a block assembly algorithm (25) is used to determine a best set of blocks representing a protein family, and the two sets are compared by the user: blocks with similar alignments obtained by the two methods may be trusted, but those that differ require scrutiny. Both Match-Box and BlockMaker require that the blocks be in order along the sequences, and so repeats might be missed. However, the EM-based program, MEME (21), does not impose an ordering criterion, and MEME finds repeats and displays them within blocks. BlockMaker, MEME and Match-Box are available from the BCM multiple alignment search launcher, which allows successive searches of a single query with several tools, both traditional and motif-based. Performance evaluation of the methods available over the Web show that there are trade-offs between sensitivity and reliability (23), and so it is worthwhile to try several methods on any particular set of sequences and compare the results.

Protocol 1

Finding motifs from unaligned sequences and searching sequence databanks

1 Go to the Swiss-Prot Random-entry retriever (http://www.expasy.ch/sprot/get-random-entry.html) and successively extract 10 Swiss-Prot sequences of length > 300 aa residues in FASTA format (look for the 'FASTA format' link at the bottom of the page).

2 Copy and paste these sequences into the large box of the BCM alignment launcher (http://dot.imgen.bcm.tmc.edu:9331/multi-align/multi-align.html). Choose BlockMaker and submit.

3 From the BlockMaker results page, examine the alignments in both sets of blocks (from Motif above and Gibbs below), and choose the set of blocks that has the most total residues. Click on the MAST direct link (to http://meme.sdsc.edu/meme/website/mast.html) above the chosen set. MAST search results will be returned by e-mail.

4 Compare the names of your submitted sequences to the significant hits from your MAST search. Other significant hits may be homologues of your submitted sequences.

5 Now that you have done the necessary control, you are ready to use your own sequences. Return to the BCM alignment launcher, copy and paste your own sequences into the box and successively click on the various choices of multiple aligners.

```
P19840 225 QLMILVNYNEDSNKAKQET..22.E IAENAVG...3.ECITAAKLA
P21707 158 LLVGIIQAAELPALDMGGT.194.DAIDKVFVG..30.EEEVDAMLA
P33232 133 DRGFMRNALERAKAAGCST..86.EWIRDFWDG.107.EMKVAMTLT
P44439 134 LRDNVYGTIEQDAARRDFT..17.EGIKDLKAG.220.EGGETIELA
P49596 100 LDQQMRVDEETKDDVSGTT.102.EFTVLACDG..28.EELLTRCLA
Q09530 676 LKAAARKRPELKLIITSAT..86.BLIILPVYG.224.EGDHLTLLA
Q56648  72 LTDVLQLPKERLLVTVYET.107.ERISAIMQG..97.ELKKQQALV
```

Figure 2 A typical example of a set of MOTIF-generated blocks obtained using Protocol 1. Boxshade was used to highlight 'conserved' positions among the randomly chosen sequences. Using these blocks as input to MAST for searching Swiss-Prot, each of the sequences represented in the block was detected with an E-value between 1.6×10^{-16} and 7.4×10^{-6}. The first unrelated sequence was detected E = 0.212.

The interactive MACAW program provides an excellent alternative to automated web-based multiple aligners. The program allows a user to choose either MACAW or Gibbs sampling for making blocks, and to make parameter choices at different stages in the alignment process. The program is available to run under popular computer operating systems.

4 Position-specific scoring matrices (PSSMs)

Alignments, patterns, logos, and trees provide useful visual displays, but for searching databases, score-based representations are most widely used. These were introduced by McLachlan (26) and popularized by Gribskov *et al.* (27) who coined the term 'position-specific scoring matrix' (or PSSM, pronounced 'possum'). A PSSM consists of columns of weights for each amino acid derived from corresponding columns of a multiple sequence alignment. Other terms have been used to describe this basic idea, including weight matrix, profile and HMM. Profiles are PSSMs constructed using the average score method (27), although the term has also been used to describe matrices representing a string of local environments for successive residues in a structure (28). Profile HMMs, described in Chapter 4, are PSSMs that are constructed using an iterative probabilistic algorithm for determination of position-specific gap penalties (29–31). A simple PSSM has as many columns as there are positions in the alignment, and 20 rows, one for each amino acid. In some applications, a PSSM consists of rows that correspond to successive positions in the alignment (27), rather than columns, and in some, there are position-specific gap scores.

Because they consist of numbers, PSSMs are useful for computer-based alignment and database searching methods but not for visual display. However, logos are computed from PSSMs, and rules can be applied to convert PSSMs to patterns (6) or consensus sequences (32). The construction of PSSMs from multiple alignments has improved over the years, and as a result, we are better able to detect weak similarities in searches (33). To construct effective PSSMs, two major issues, described below, must be addressed.

4.1 Sequence weights

PSSM performance can be improved by differentially weighting sequences to reduce redundancy resulting from non-representative sampling of sequences (34–37). Very similar sequences get low weights and more diverged sequences get higher weights in order to make a PSSM more representative of the family as a whole. Several different strategies have been used to arrive at sequence weights. Some methods start with a tree and find a root, where the weight of a sequence is proportional to its distance from the root (e.g. 35). Pairwise distance methods calculate a weight from the average distance of a sequence to all other sequences, either to the observed sequences or to imaginary sequences derived by sampling residues from the observed sequences (e.g. 36). Position-based sequence weights are calculated by determining the weight of a residue within its column in an alignment and adding residue weights for all positions (37). In the maximum discrimination method, weights are chosen to best discriminate between true positives and background (31). Comprehensive empirical evaluation of sequence weighting methods has revealed that weighting sequences is much better than not weighting at all (37). However, no single method stands out, and at least one variant of each of these strategies provided excellent results.

4.2 PSSM column scores

Pairwise alignment methods utilize amino acid substitution matrices to provide a set of scores for each aligned residue. Current applications utilize log-odds scores computed from alignment data, such as PAM (38), JTT (39), or BLOSUM (40) substitution matrices, and theory advocates the suitability of log-odds scores for pairwise alignments (41). However, scoring a multiple alignment against a sequence is more complex, requiring a scoring scheme that is able to utilize the observed occurrences of residues in a column corresponding to an alignment position. That is, the column of an alignment should be modelled in a way that, when aligned with each of the 20 amino acids, a meaningful score can be obtained. The original average score method (27) simply extends the use of pairwise scores by averaging them. For instance, when aligned with a serine, a position represented by an alanine and 3 cysteines would get a score equal to the serine–alanine score plus 3 times the serine–cysteine score divided by 4 (ignoring sequence weights which would alter the relative contribution of each occurrence to the sum).

Unfortunately the average score is insensitive to the number of sequences in the multiple alignment. The average score for a serine aligned with an alanine and 3 cysteines is identical to that for a serine aligned with 10 alanines and 30 cysteines. The problem with this situation is that a serine might be expected to occur frequently given only 4 observations of such similar residues, but after 40 observations without seeing a serine, we would expect to see one only rarely. Therefore, the average score method becomes less and less realistic as the number of different sequences in an alignment increases. An effective way of dealing with this problem is to add 'pseudocounts' to the observed counts of residue occurrences (42–44). Intuitively, this is equivalent to adding hypothetical sequences to those that have been observed, and for each sequence, the choice of residues at each aligned position is governed by what might be expected for real related sequences not yet seen. So if we have already observed an alanine and 3 cysteines, we might expect to see more cysteines and alanines, but also occasional serines but maybe not arginines. Hypothetical occurrences can be added to real occurrences as fractional pseudocounts. Notice that if we add the same pseudocounts when 10 alanines and 30 cysteines have been seen, then the relative proportion of real observations to pseudocounts increases 10-fold; this conforms with our intuition that we are much more certain that the observed occurrences adequately model future occurrences when we have a large number of independent observations. Comprehensive evaluations demonstrate that using pseudocounts modeled on alignment data, much better overall performance is obtained than using the average score method (33, 44).

5 Searching family databases with sequence queries

For any protein sequence of interest, a search of the latest databanks is the first and often the most important step toward understanding function, and the identification of homologues in this way has been a major driving force in both academic biology and in the growing genomics industry. A second step should involve searching protein family databases. There are several reasons for this: Making sense of dozens or hundreds of hits in the sequence databanks can be challenging, whereas hits in protein family databases provide immediate classification and entries to the literature. The different regions of multi-domain proteins are readily classified using family databases, whereas in searches of sequence databanks, modules can be missed if hits to family members containing them are low on the list. Searches of family databases can be more sensitive than searches of sequence databanks because multiple alignment information is utilized. The much smaller size of family databases, typically only ~1% the size of sequence databanks, reduces noise.

Currently, there are several choices of family databases and searching options available over the internet (*Table 1*). An illustrative example is depicted in *Figure 3*, which shows how well the different methods detected key features of a protein that we recently described, a cytosine-5 DNA methyltransferase homologue with an embedded chromodomain module, called a 'chromomethylase', which is encoded by the *Arabidopsis thaliana* CMT1 locus (45). In addition to being the subject of current experimental work in our group, we chose this sequence because chromomethylases are not yet present in any of the family databases, although both the cytosine-5 DNA methyltransferases and the chromodomains are represented in all of them, and because this example reveals strengths and weaknesses of the different methods especially well. Both the DNA methyltransferase and the chromodomain represent novel subfamilies of their respective families, and so detection in their entirety can be challenging for a protein family classification method that does not generalize well from known examples. This is an anecdotal example, and overall performance can only be judged using

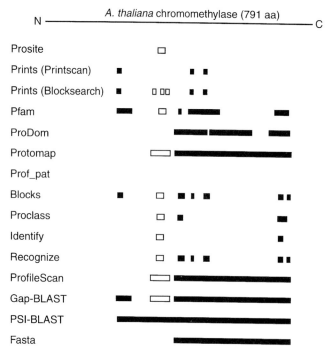

Figure 3 Classification of the 791 aa *A. thaliana* chromomethylase by family databases. The horizontal line indicates the length of the protein from the amino (N) to the carboxyl (C) end, the closed boxes show the extent of cytosine-5 DNA methyltransferase regions detected and the open boxes show chromodomain regions. For methods that report E- or p=values, a 0.05 level of significance was considered to be the threshold for detection, and this exceeded the level of the highest-scoring false positives. For methods that do not report E- or p-value statistics (Prosite, Printscan, Prof_pat, Proclass) or for those that report multiple levels of stringency (Identify and Recognize), the threshold level of detection was considered to be just above the first false positive hit.

comprehensive empirical evaluations. However, because coverage of different databases varies widely, such rigorous direct comparisons have not been carried out.

5.1 Curated family databases: Prosite, Prints, and Pfam

Prosite is the original family database, introduced in 1989. Prosite provides excellent documentation and carefully crafted patterns for searching (*Section 2.2*). In cases where patterns are difficult to find, Prosite provides a profile PSSM (*Section 4*). Prosite 15 (July, 1998) has 1020 documentation entries, mostly representing families, and 1358 patterns based on sequences in Swiss-Prot. Searching a query sequence against Prosite patterns is strictly a hit-or-miss affair, and no statistics are provided. The chromomethylase example illustrates this vividly. Prosite reported the chromodomain, which is a highly diverged module that is relatively difficult to detect. However, Prosite failed to detect the DNA methyltransferase, even though some of the conserved regions are very easily detected by standard searching programs (*Figure 3*) and this family is represented by two patterns in the database. Indeed, a comprehensive empirical evaluation showed that even standard BLAST searching outperforms searching of Prosite patterns (46).

Prints, introduced in 1993, is similar to Prosite in providing excellent documentation. Rather than patterns, Prints provides carefully crafted 'fingerprint' multiple alignments (ordered sets of blocks), that can be searched using pattern or PSSM methods. Prints 20 (October, 1998) has 990 fingerprint entries for 5701 blocks based on sequences in the OWL protein database. Printscan detected all 3 blocks in the fingerprint representing the upstream and central conserved regions of the DNA methyltransferase, but did not detect the chromodomain.

Maintaining curated databases and crafting patterns or fingerprints is made especially difficult because of the rapid expansion of protein families in recent years. Pfam (47), introduced in 1996, addresses this problem by using seed alignments that are manually constructed, and HMM (hidden Markov model) PSSMs from the seeds are then used to automatically extract and align new sequences from databanks. Unlike Prosite and Prints, Pfam does not provide documentation beyond a family name and links to source databases and does not delineate conserved regions within entries. Pfam 3.2 (October, 1998) has 1344 entries representing families and modules based on sequences in Swiss-Prot/TrEMBL. HMM PSSMs are used to search Pfam. For the chromomethylase, all of the conserved regions of the DNA methyltransferase and the chromodomain were detected.

5.2 Clustering databases: ProDom, DOMO, Protomap, and Prof_pat

An alternative to curation is to search a database against itself, then cluster similar sequences into families automatically. Although the procedure sounds simple, in practice it is fraught with difficulties owing to the complexity of proteins and protein families and to the need to avoid chance similarities when comparisons are carried out on such a large scale. The first public database of this type was introduced in 1990 (48), and several have been introduced over the years, only some of which are extant. ProDom, which was introduced in 1994 (49), has been continually maintained and enhanced (50); version 36 (August, 1998) contains 17 777 entries from Swiss-Prot with more than 2 sequences. ProDom entries vary from short single motifs to longer stretches of similarity that might encompass nearly entire sequences. ProDom is searched with multiple alignments or consensus sequences. Using either option, ProDom detected the central and downstream conserved regions of the DNA methyltransferase, missing the upstream region and the chromodomain.

Recently, three new clustering databases have been introduced. DOMO (51), which is based on Swiss-Prot and PIR, is similar to ProDom, although it uses different methodology to generate the database. DOMO clusters tend to be longer and fewer in number than ProDom clusters. At present, DOMO does not allow user-supplied sequences to be searched for classification. Protomap (52), which is based on Swiss-Prot, does not yield multiple alignments as do ProDom and DOMO, but rather provides a graphical tree-like view of the clustering. To classify a protein sequence with Protomap, a Smith–Waterman search of Swiss-Prot is performed, and each individual cluster that contains a sequence hit is reported. For the chromomethylase, Protomap detected the chromodomain and the central and downstream conserved regions of the DNA methyltransferase, missing the upstream region of conservation. Prof_pat (53) extracts patterns from clustering Swiss-Prot/TrEMBL, and these can be searched. Prof_pat did not detect either the DNA methyltransferase or the chromodomain above false positives.

5.3 Derived family databases: Blocks and Proclass

Intermediate between the curated and automated databases are those that utilize protein family groupings provided by other resources. The Blocks Database, which was introduced in 1991, uses the automated Protomat system for finding blocks (ungapped regions of local conservation) representing a protein family. Starting with Swiss-Prot sequences listed in Prosite family entries, alignment blocks are found (patterns or profiles provided with Prosite are not used) and concatenated into a database. Blocks 11.0 (August, 1998) contains 994 families and 4034 blocks based on Swiss-Prot and is searched using the PSSM-based BlockSearch method that reports single and multiple block hits along a sequence. Whereas other protein family searchers on the internet require a protein sequence, Blocks can be searched with a DNA sequence query, in which case hits from all three frames on each strand are assembled. For searching, the current default database is Blocks+, a superset of families from Blocks, Prints, Pfam, ProDom and DOMO. Blocks+ (Nov. 1998) includes 8388 blocks representing 1922 families. Except for Prints, where fingerprint blocks are utilized directly for searching, Protomat is used to make blocks for entries from each database, and families that have block regions in common are removed to avoid redundancy. BlockSearch detected the chromodomain and all of the conserved regions of the DNA methyltransferase. When the Prints database was searched with BlockSearch, all 3 upstream and central DNA methyltransferase motifs and all 3 chromodomain motifs in Prints were now detected at highly significant levels.

Proclass (54), introduced in 1997, also combines families from different sources: Prosite, PIR superfamilies and families automatically discovered using the GenFind program (55). Proclass v. 3 (March, 1998) contains 1275 Prosite groups and 3979 PIR superfamilies and is searched using a neural network-based system. To our knowledge, Proclass searching is the only system that detects sequence similarity using methodology that is not alignment-based. When the chromomethylase sequence was searched, Proclass reported the Prosite chromodomain pattern and both of the Prosite DNA methyltransferase patterns, which were missed by the Prosite scanner.

5.4 Other tools for searching family databases

Identify (7) searches sequences versus pattern-based representations of individual blocks and fingerprints derived from the Blocks and Prints databases. Because patterns can be searched much more rapidly than scored-based representations of multiple alignments (see Chapter 7), Identify search results are returned within a second or so. Identify detected the chromodomain and only one down-stream DNA methyltransferase blocks above all false positives. Using Recognize, which is a score-based version of Identify, the central and other downstream regions were detected as well.

A collection of profile PSSMs from Prosite, Pfam and other sources is available for searching using generalized HMM-like profile PSSMs at the ProfileScan site (56). ProfileScan reported the chromodomain and the central and downstream DNA methyltransferase conserved regions but missed the conserved upstream region.

In summary, there are numerous protein family searching tools available for sequence classification. None is perfect, and as illustrated by the chromomethy-lase example, it is worthwhile to try several of them for analysing a sequence of interest. Pairwise sequence tools also varied in their ability to confidently detect features of the chromomethylase. GAP-BLAST detected the chromodomain and all DNA methyltransferase conserved regions, although subsequent iterations of PSI-BLAST caused the chromodomain to be lost at the expense of the DNA methyltransferase that surrounds it. FASTA failed to detect the upstream con-served region of the DNA methyltransferase, and the chromodomain was re-ported, but at a non-significant level. As a practical matter, the chromodomain would have gone unnoticed or assumed to be a chance hit because it is preceded by ~100 higher-scoring DNA methyltransferase sequences in databanks, and indeed its presence was not noted in the original sequence entry (GenBank/EMBL U53501). By using family databases for classification, this potential problem can be minimized.

6 Searching with family-based queries

Finding homologues in sequence databanks underlies much of the recent progress in functional genomics, both in academia and industry, and pairwise methods, such as BLAST searching, currently dominate. However, as more and more sequences fall into families, opportunities increase for using family information for identifying modules and new family members. Progress in making better PSSMs described in *Section 3.3* has resulted in improvements in searching performance, and practical tools have become available for taking advantage of protein family information in searching sequence databanks.

6.1 Searching with embedded queries

A potential drawback to block-based approaches is that regions of uncertain alignment are not scored, and the loss of this alignment information can potentially reduce searching sensitivity. This problem arises because even with effective motif-finding systems, the 'edges' of blocks are often uncertain, and they might be chosen differently for different subsets of proteins in a family (51, 57, 58). This problem has been addressed by implementation of a simple 'embedding' strategy: a consensus is determined for a set of related sequences, the sequence that is closest to the consensus is chosen, and blocks are embedded into that sequence (46). Because interblock regions of uncertain alignment are represented as a single sequence, they cannot be misaligned (this would reduce the specificity of a PSSM), while multiple alignment information in block regions is retained. Embedding of PSSMs using this system has not been implemented on the internet for general database searching, although the basic idea has been incorporated into PSI-BLAST (described below). As an approximation, using the COBBLER (COnsensus Biasing By Locally Embedding Residues) system, a consensus residue is determined for each position of all the blocks. A single sequence is chosen as the one closest to the consensus over all block positions, and these consensus residues are then substituted for the real residues in the chosen sequence. This consensus-biased sequence can then be used to search sequence databanks using available single sequence querying tools, such as BLAST and PSI-BLAST. The improved overall performance that results is especially useful for identifying known modules in unexpected places: For instance, the chromodomain in the *A. thaliana* chromomethylase (*Figure 3*) was initially identified using a COBBLER-embedded sequence to search the nr protein databank with BLAST (45).

6.2 Searching with PSSMs

Using PSSMs to search sequence databanks is computationally demanding, and the availability of services is relatively limited. The Multiple Alignment Searching Tool (MAST) program (57) searches block-based multiple alignments against the standard sequence database sets, which are updated daily. MAST output provides excellent statistics for both individual and multiple block hits with block maps for intelligent interpretation of search results. MAST accepts PSSMs directly from MEME and BlockMaker. Additionally, the Blocks server provides a processor that can be used to convert other multiple alignments into efficient PSSMs for sending directly to the MAST server.

6.3 Iterated PSSM searching

Several of the concepts highlighted above have been incorporated into the Gap/PSI-BLAST searcher, an elegant extension of the popular BLAST database searching program (59). The first round of searching employs Gap-BLAST, a new pairwise method for detecting family members in the traditional way. From the significant hits detected in the first round, a PSSM is constructed and this is used to search the databank again, a process that can be repeated multiple times until no further hits are reported above a chosen level of significance.

Gap-BLAST is especially notable because it represents a successful block-based approach to the pairwise searching problem. When databanks are searched, computational speed is an important factor, because finding an 'optimal' alignment using traditional methods for placing gaps is too slow to be practical on a large scale with standard hardware. Speedy methods, such as FASTA and BLAST, begin by searching exhaustively for matches or short motifs shared by two sequences, extending these and stringing them together to find high scoring alignments. However, the gain in speed is accomplished at the expense of reduced searching performance (60). Gap-BLAST combines speed with near-optimal searching performance by starting with short motifs, but accepting only those that define opposite ends of a high scoring ungapped alignment. This alignment is extended, and only if it exceeds a threshold score is a gapped alignment sought, that is, gapping is employed to optimize alignment of highly similar regions. Searching performance of Gap-BLAST is nearly indistinguishable from that of an optimal gap placement method (Smith–Waterman dynamic programming) when the same scoring parameters are used. This is one inroad of blocker concepts into the gapper realm for pairwise alignment; another is the realization that pairwise alignment can be generally improved by allowing highly dissimilar regions to be skipped over (61, 62).

PSSM construction in PSI-BLAST is similar to that described in *Section 4*, employing position-based sequence weights (37) and pseudocounts that are modelled upon amino acid substitution probabilities (33, 44). The embedding concept described above is generalized in PSI-BLAST to deal with the complication that for any position in the query sequences, there may be a variable number of database sequences that align. Thus, the final PSSM provides position-specific scores that represent as few as one (the query sequence alone) and as many as all of the sequences detected in the previous round of the search. The high sensitivity to distant relationships provided by PSI-BLAST, and the enjoyment that a user may get by iteratively searching for homologues in real time, can lead to its overenthusiastic use, and serious errors may result. This is because any chance hit that is included in the developing PSI-BLAST PSSM will almost inevitably pull out its neighbours in subsequent rounds, and this can lead to erroneous inferences of homology. A defence against this type of error is to use conservative levels of statistical significance for addition of sequences to the PSSM. However, because proteins are not comprised of random sequences of residues, the random statistical model that underlies the BLAST programs can be unreliable (63), and so novel conclusions drawn from iterative searches should be viewed with appropriate caution.

6.4 Multiple alignment-based searching of protein family databases

The effectiveness of utilizing family-based information to search databases encourages the use of multiple alignments for searching multiple alignment databases. LAMA (for local alignment of multiple alignments) is a program that searches ungapped blocks versus family databases (64). In LAMA, PSSM columns are scored against one another by calculating a correlation coefficient, and a high scoring alignment is one in which the sequence-weighted distribution of residues is highly similar overall between aligned columns. The high sensitivity of LAMA for locally aligned regions has led to its use in discovering subtle similarities, such as those shared by helix-turn-helix DNA binding motifs found in unrelated modules.

Tools such as LAMA, which thrive on abundant alignment data, are likely to become more widely used as protein families expand in size. Because of its high sensitivity, LAMA or its descendants should become increasing valuable for modelling 3-D structures of sequences by facilitating local alignment to family members of known structure. As the percentage of unclassified proteins dwindles, a major alignment-based problem facing biologists will be to determine which subfamily a protein belongs to, and from this, more precise structural and functional inferences may be made. We anticipate the development of a next generation of computational tools to deal with this problem.

References

1. Rost, B. (1996). In *Methods in enzymology* (ed. R. F. Doolittle), Vol. 266, p. 525. Academic Press.
2. Garnier, J., Gibrat, J.-F., and Robson, B. (1996). In *Methods in enzymology*, Vol. 266, p. 540.
3. Moult, J., Hubbard, T., Bryant, S. H., Fidelis, K., and Pedersen, J. T. (1997). *Proteins: Struct. Funct. Genet.*, **Suppl. 1**, 2.
4. Henikoff, S. (1991). *New Biol.*, **3**, 1148.
5. Bairoch, A., Bucher, P., and Hofmann, K. (1997). *Nucleic Acids Res.*, **25**, 217.
6. Jonassen, I., Collins, J. F., and Higgins, D. G. (1995). *Protein Sci.*, **4**, 1587.
7. Nevill-Manning, C. G., Wu, T. D., and Brutlag, D. L. (1998). *Proc. Natl. Acad. Sci. USA*, **95**, 5865.
8. Schneider, T. D. and Stephens, R. M. (1990). *Nucleic Acids Res.*, **18**, 6097.
9. Saitou, N. and Nei, M. (1987). *Mol. Biol. Evol.*, **4**, 406.
10. Vingron, M. and Argos, P. (1991). *J. Mol. Biol.*, **218**, 33.

11. Boguski, M. S., Hardison, R. C., Schwartz, S., and Miller, W. (1991). *New Biol.*, **4**, 247.
12. Schuler, G. D., Altschul, S. F., and Lipman, D. J. (1991). *Proteins. Struct. Funct. Genet.*, **9**, 180.
13. Sobel, E. and Martinez, H. M. (1986). *Nucleic Acids Res.*, **14**, 363.
14. Posfai, J., Bhagwat, A. S., Posfai, G., and Roberts, R. J. (1989). *Nucleic Acids Res.*, **17**, 2421.
15. Smith, H. O., Annau, T. M., and Chandrasegaran, S. (1990). *Proc. Natl. Acad. Sci. USA*, **87**, 826.
16. Depiereux, E. and Feytmans, E. (1992). *CABIOS*, **8**, 501.
17. Neuwald, A. F. and Green, P. (1994). *J. Mol. Biol.*, **239**, 698.
18. Bacon, D. J. and Anderson, W. F. (1986). *J. Mol. Biol.*, **191**, 153.
19. Stormo, G. D. and Hartzell, G. W. 3rd (1989). *Proc. Natl. Acad. Sci. USA*, **86**, 1183.
20. Lawrence, C. E., Altschul, S. F., Boguski, M. S., Liu, J. S., Neuwald, A. F., and Wootton, J. C. (1993). *Science*, **262**, 208.
21. Bailey, T. and Elkan, C. (1994). In *Proceedings of the Second International Conference on Intelligent Systems for Molecular Biology*, pp. 28–36. AAAI Press, Menlo Park, CA.
22. Notredame, C., Holm, L., and Higgins, D. G. (1998). *Bioinformatics*, **14**, 407.
23. Briffeuil, P., Baudoux, G., Lambert, C., De Bolle, X., Vinals, C., Feytmans, E., *et al.* (1998). *Bioinformatics*, **14**, 357.
24. Henikoff, S., Henikoff, J. G., Alford, W. J., and Pietrokovski, S. (1995). *Gene*, **163**, GC17.
25. Henikoff, S. and Henikoff, J. G. (1991). *Nucleic Acids Res.*, **19**, 6565.
26. McLachlan, A. D. (1983). *J. Mol. Biol.*, **169**, 15.
27. Gribskov, M., McLachlan, A. D., and Eisenberg, D. (1987). *Proc. Natl. Acad. Sci. USA*, **84**, 4355.
28. Bowie, J. U., Luthy, R., and Eisenberg, D. (1991). *Science*, **253**, 164.
29. Krogh, A., Brown, M., Mian, I. S., Sjolander, K., and Haussler, D. (1994). *J. Mol. Biol.*, **235**, 1501.
30. Baldi, P., Chauvin, Y., Hunkapiller, T., and McClure, M. A. (1994). *Proc. Natl. Acad. Sci. USA*, **91**, 1059.
31. Eddy, S. R., Mitchison, G., and Durbin, R. (1995). *J. Comput. Biol.*, **2**, 9.
32. Patthy, L. (1987). *J. Mol. Biol.*, **198**, 567.
33. Henikoff, J. G. and Henikoff, S. (1996). *CABIOS*, **12**, 135.
34. Luthy, R., Xenarios, I., and Bucher, P. (1994). *Protein Sci.*, **3**, 139.
35. Thompson, J. D., Higgins, D. G., and Gibson, T. J. (1994). *CABIOS*, **10**, 19.
36. Sibbald, P. R. and Argos, P. (1990). *J. Mol. Biol.*, **216**, 813.
37. Henikoff, S. and Henikoff, J. G. (1994). *J. Mol. Biol.*, **243**, 574.
38. Dayhoff, M. (1978). *Atlas of protein sequence and structure*, Vol. 5, suppl. 3, pp. 345–58. National Biomedical Research Foundation, Washington, DC.
39. Jones, D. T., Taylor, W. R., and Thornton, J. M. (1992). *CABIOS*, **8**, 275.
40. Henikoff, S. and Henikoff, J. G. (1992). *Proc. Natl. Acad. Sci. USA*, **89**, 10915.
41. Altschul, S. F. (1991). *J. Mol. Biol.*, **219**, 555.
42. Dodd, I. B. and Egan, J. B. (1987). *J. Mol. Biol.*, **194**, 557.
43. Brown, M. P., Hughey, R., Krogh, A., Mian, I. S., Sjolander, K., and Haussler, D. (1993). In *Proc. First Int. Conf. on Intelligent Systems for Molecular Biology* (ed. L. Hunter, D. Searls, and J. Shavlik), pp. 47–55. AAAI Press, Washington DC.
44. Tatusov, R. L., Altschul, S. F., and Koonin, E. V. (1994). *Proc. Natl. Acad. Sci. USA*, **91**, 12091.
45. Henikoff, S. and Comai, L. (1998). *Genetics*, **149**, 307.
46. Henikoff, S. and Henikoff, J. G. (1997). *Protein Sci.*, **6**, 698.
47. Sonnhammer, E. L., Eddy, S. R., and Durbin, R. (1997). *Proteins: Struct. Funct. Genet.*, **28**, 405.

48. Smith, R. F. and Smith, T. F. (1990). *Proc. Natl. Acad. Sci. USA*, **87**, 118.

49. Sonnhammer, E. L. L. and Kahn, D. (1994). *Protein Sci.*, **3**, 482.

50. Corpet, F., Gouzy, J., and Kahn, D. (1998). *Nucleic Acids Res.*, **26**, 323.

51. Gracy, J. and Argos, P. (1998). *Bioinformatics*, **14**, 174.

52. Gona, G., LInial, N., Tishby, N., and Linial, M. (1998). *ISMB*, **6**, 212.

53. Bachinsky, A. G., Yarigin, A. A., Guseva, E. H., Kulichkov, V. A., and Nizolenko, L. P. (1997). *CABIOS*, **13**, 115.

54. Wu, C. H., Zhao, S., and Chen, H. L. (1996). *J. Comput. Biol.*, **3**, 547.

55. Wu, C. H., Shivakumar, S., Shivakumar, C. V., and Chen, S. C. (1998). *Bioinformatics*, **14**, 223.

56. Bucher, P., Karplus, K., Moeri, N., and Hofmann, K. (1996). *Comput. Chem.*, **20**, 3.

57. Bailey, T. L. and Gribskov, M. (1997). *J. Comput. Biol.*, **4**, 45.

58. Nicodeme, P. (1998). *Bioinformatics*, **14**, 508.

59. Altschul, S. F., Madden, T. L., Schaffer, A. A., Zhang, J., Aneng, Z., Miller, W., *et al.* (1997). *Nucleic Acids Res.*, **25**, 3389.

60. Pearson, W. R. (1995). *Protein Sci.*, **4**, 1145.

61. Altschul, S. F. (1998). *Proteins: Struct. Funct. Genet.*, **32**, 88.

62. Alexandrov, M. M. and Luethy, R. (1998). *Protein Sci.*, **7**, 254.

63. Brenner, S. E., Chothia, C., and Hubbard, T. J. (1998). *Proc. Natl. Acad. Sci. USA*, **95**, 6073.

64. Pietrokovski, S. (1996). *Nucleic Acids Res.*, **24**, 3836.

Chapter 6

Predicting secondary structure from protein sequences

Jaap Heringa

National Institute for Medical Research, The Ridgeway, Mill Hill, London NW7 1AA, UK

1 Introduction

Protein structure is intrinsically hierarchic in its internal organization. The highest level in this hierarchy is constituted by complete proteins or assemblies of such proteins, which become subdivided through domains via super-secondary structure to secondary structure at the lowest hierarchical level.

At higher levels within in this hierarchy, especially from the domain level upwards, the connectivity of the polypeptide backbone between substructures becomes less important. A protein thus can retain a stable structure irrespective of the sequential arrangement of domains and presence of fragments linking them together. Such linker regions often constitute exposed surface loops that do not disrupt the folds of the domains they connect (1).

At the level of protein secondary structure, however, the elements are not only crucially dependent on their amino acid compositions, but, unlike domain and higher-order structures, are also very much context dependent; i.e. they rely critically on the substructures in their environment. It is because of this context dependency, that predicting protein secondary structure is a very difficult task, which after three decades of research has not attained the accuracy on which further prediction of tertiary structure can be based. It must be stressed, however, that some successful predictions of higher-order structure, based on a knowledge of the secondary structure, have been achieved (e.g. ref. 2).

This chapter covers some background aspects of secondary structure prediction and describes recent and successful prediction methods, most of which are available through the World Wide Web and so can be used by virtually every biologist who likes to find out about the secondary structure associated with a particular protein query sequence.

1.1 What is secondary structure?

Perhaps a suitable definition in the context of this chapter for a secondary structure is that it is a consecutive fragment in a protein sequence, which corresponds to a local region in the associated protein structure showing

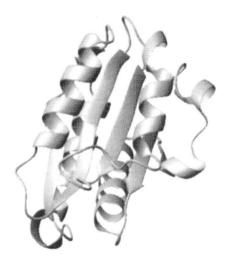

Figure 1 Ribbon diagram of the flavodoxin fold (PDB code 5nul) belonging to the α/β doubly-wound fold family.

distinct geometrical features. The two basic secondary structures are the α-helix and the β-strand. Both show distinct structural features and are easily recognizable in a protein structure (*Figure 1*). Other secondary structure types occurring in protein structures are more difficult to classify as they are less regular than α-helices or β-strands. Such structures are defined in the context of most prediction methods as coil; i.e. leftover secondary structures that cannot be considered in the α-helical or β-stranded conformation.

In general, about 50% of the amino acids fold into α-helices or β-strands, so that roughly half the protein structures are irregularly shaped. The primary reason for the regularity observed for helices and strands is the inherent polar nature of the protein backbone, which contributes a polar nitrogen and oxygen atom for each amino acid. To satisfy energetical constraints, the parts of the main-chain buried in the internal protein core need to form hydrogen-bonds between those polar atoms. The α-helix and β-strand conformations are optimal as each main-chain nitrogen atom can associate with an oxygen partner (and *vice versa*) whenever they adopt one of these two secondary structure types. It must be stressed that, in order to satisfy their hydrogen-bonding constraints, β-strands need to interact with other β-strands, which they can do in a parallel and anti-parallel fashion, thus forming a β-pleated sheet. β-strands thus depend on crucial long-range interactions between residues remote in sequence. They therefore are more context dependent than α-helices, which would be more able to fold 'on their own'. The fact that the vast majority of prediction methods have greatest difficulty in delineating β-strands correctly is believed to be due to their pronounced context dependency.

1.2 Where could knowledge about secondary structure help?

Experimental evidence on early protein folding intermediates has shown that secondary structural elements form at early stages during the folding process (for a review, see ref. 3). These results support the significance of the so-called 'framework' model of protein folding (4, 5), where two or more secondary structural elements would associate early during folding to provide a structural frame to which subsequently other substructures could attach. Therefore, knowledge of protein secondary structural regions along the sequence is a prerequisite to model the folding process or kinetics associated with it. Also for tertiary model building, the ability to predict the secondary structure from the sequence alone is crucial, as it allows for docking experiments to be carried out on the predicted α-helices and β-strands.

On the architectural side of protein structure, it is possible to recognize the three-dimensional topology by comparing the successfully predicted secondary structural elements of a query protein with a database of known topologies. Successful prediction here means parts of those helices and strands essential for the topology would have been predicted, without necessarily accurate prediction of the edges of those structures or the detection of non-essential secondary structures. An example of topologically essential secondary structures for the flavodoxin fold is given in *Figure 2*. The figure shows a schematic representation

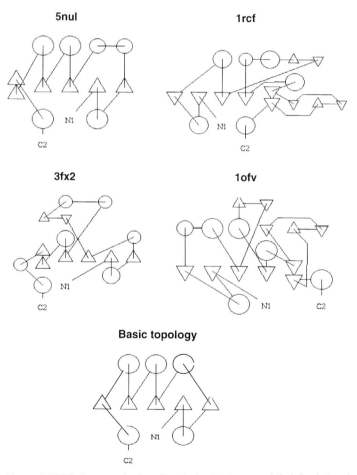

Figure 2 TOPS diagrams for four flavodoxin structures and their basic topology. The essential secondary structures are given in the basic topology diagram.

115

of the secondary structures as provided by the TOPS server (http://tops.ebi.ac.uk). In a TOPS diagram (6), a α-helix is represented by a circle and a β-strand by a triangle. The flavodoxin fold belongs to the class of α/β-folds with the essential secondary structures distributed over the sequence as $[\beta\alpha]_5$. The five strands fold into a single β-pleated sheet ordered topologically as β_2-β_1-β_3-β_4-β_5, where the numbers indicate their relative position in the sequence and hyphens the hydrogen bonded and spatial interactions between the strands. The five α-helices, each following a β-strand, shield the β-sheet from the solvent and therefore are of an amphipathic nature (see below). From the topologies of a few different flavodoxin structures (*Figure 2*) can be seen that varying substructures can be added on to the basic structure, albeit they do not disrupt the fold of the topologically essential secondary structures. Therefore, proper prediction of the sequential order of the topologically essential helices and sheets often allows the recognition of the fold type associated with the protein sequence con-sidered, thereby conferring the information pertaining to that fold. Further-more, active sites of enzymes typically comprise amino acids positioned in loops, so that, for example, identically conserved residues at multiple alignment sites predicted to be in loop regions (i.e. not predicted as α-helix or β-strand), could be functional and together elucidate the function of the protein (or protein family) under scrutiny.

1.3 What signals are there to be recognized?

A number of observations on secondary structures as found in the large collec-tion of protein structures deposited in the Protein Data Bank (PDB) (7), could be summed up for each of the secondary structures α-helix, β-strand, and loop as follows.

α-helix:

(a) As the number of residues per turn is 3.6 in the ideal case and helices are often positioned against a buried core, they have one phase contacting hydro-phobic amino acids, while the other phase interacts with the solvent. Such amphipathic helices (8) thus show a periodicity of three to four residues in hydrophobicity of the associated sequence stretch (*Figure 3*).

(b) Proline residues do not occur in middle segments as they disrupt the α-helical turn. However, they are seen in the first two positions of α-helices.

β-strand:

(a) β-Strands mostly fold into so called β-pleated sheets which have two strands forming either edge. Therefore the hydrophobic nature of edge strands is dif-ferent from that of strands internal to a β-sheet. As side-chains of constituent residues along a β-strand alternate the direction in which they protrude, edge strands of β-sheets can show an alternating pattern of hydrophobic-hydrophilic residues, while buried strands tend to contain merely hydro-phobic residues (*Figure 3*).

(b) As β-strand is the most extended conformation (i.e. consecutive Cα atoms are farthest apart), it takes relatively few residues to cross the protein core with a strand. Therefore, the number of residues in a β-strand is usually limited and can be anything from two or three amino acids, whereas helices shielding such strands from solvent comprise more residues.

(c) The β-strands can be disrupted by single residues that induce a kink in the extended structure of the main-chain. Such so-called β-bulges are often comprised of relatively hydrophobic residues.

Coil:

(a) Multiple alignments of protein sequences often display gapped and/or highly variable regions, which would be expected to be associated with loop regions rather than the two basic secondary structures.

(b) Loop regions contain a high proportion of small polar residues like Ala, Gly, Ser, and Thr. Glycine residues are seen in loop regions due also to their inherent flexibility.

(c) Proline residues are often seen in loops as well. They are not observed in helices and strands as they kink the main-chain, although they can occur in the N-terminal two positions of α-helices as mentioned above.

In addition to the positional requirements in hydrophobicity, there are also general compositional differences between helix, strand and coil conformations and this is the signal used in many of the early prediction methods (see below) for single sequences. Methods that utilize multiple alignments can also exploit the fact that the amino acid exchange patterns are different for the three secondary structure states.

A few additional rules can help in clarifying the structure or function of a protein sequence, once the secondary structure is predicted:

(a) Hydrophobic and particularly conserved hydrophobic residues are normally buried in the protein core.

(b) More than 95% of all so-called β-α-β motifs; i.e. a β-strand followed in sequence by a α-helix and another β-strand, show a right-handed chirality. The afore-mentioned flavodoxin family (*Figure 2*) indeed shows only right-handed β-α-β motifs. This fact can be used to build a topology for the secondary structures of the sequence(s) considered.

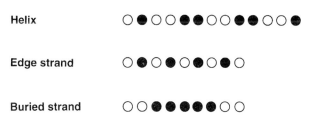

Figure 3 Hydrophobic patterns along secondary structures.

(c) Helices often cover up a core of β-strands. Therefore, if both α-helices and β-strands are predicted, an attempt should be made to distribute the helices evenly at either phase of a tentative β-sheet in topology modelling.

(d) As mentioned, strictly conserved residues in different regions of a multiple alignment can be predicted with great confidence to be responsible for the catalytic functions, particularly if they are polar and predicted to be in loop structures hence unlikely to be buried. As active site residues are positioned together in a protein 3-D structure, the coil structures they constitute should be brought together in a topology model.

2 Assessing prediction accuracy

The most widely used way to assess the quality of an alignment is by calculating the overall per residue three-state accuracy, called the Q_3:

$$Q_3 = [(PH + PE + PC)/N] \times 100\%,$$

where N is the total number of residues predicted and PS is the number of correctly predicted residues in state S ($S = H$, E, or C). Some researchers use the so-called Matthews' correlation coefficient as it more stringently estimates the prediction accuracy for each structural state:

$$C_S = \frac{(P_S + N_S) - (\sim P_S \times N_S)}{\sqrt{(P_S + \sim P_S) \times (P_S + \sim N_S) \times (N_S + \sim P_S) \times (N_S + \sim N_S)}}$$

where P_S and N_S are respectively the number of positive and negative cases correctly predicted for the structural state considered, and $\sim P_S$ and $\sim N_S$ the numbers of false positives and negatives, respectively. Three-state predictions would thus yield three Matthews' correlation coefficients. If overprediction or underprediction occurs for any of the structural states, this is more dramatically reflected in the Matthews' correlations than in the Q_3 percentage. A third way to assess prediction accuracy is by *weights of evidence*, defined for each secondary structural type S as:

$$W_S = \log[(P_S \times N_S) / (\sim P_S \times \sim N_S)].$$

Although this measure is relatively robust to different sampling frequencies of the structural states, the interpretation of the resulting values is not as straightforward as for the other two measures. Because understanding the Q_3 measure is the easiest and its use leads to just one percentage, it is the measure most frequently used in the literature to report prediction accuracy.

A very important issue in assessing performance is the notion of sustained accuracy. Knowledge about the average accuracy of a given method over a set of predicted proteins is not meaningful if unaccompanied by the variance of those predictions. It is important to know what worse case predictions can be expected from a method, even if its mean accuracy is quite high.

A standard scenario to assess prediction accuracy is the *jackknife* test carried out over a large set of test proteins (see *Protocol 1*). This ensures that no information about a query sequence or multiple alignment is used in training the method. Nonetheless, unnoticed but systematic tuning of the method to the database might still occur, so that the most rigorous test of any method is the prediction of test cases that have no homologues in the database and have not been seen during the development of the method.

Notwithstanding the importance of the measures for accuracy as listed above, the real success of a secondary structure prediction depends on how the knowledge is being used. An example is the aforementioned fold recognition, where correct prediction of the edges of secondary structural elements is not essential, but missing structures that are crucial for the basic topology is costly. However, all above measures, equally penalize, for example, missing two residues at either side of a seven-amino acid strand or missing a complete topologically essential strand of four residues.

Protocol 1

Jackknife testing

1 Take out one protein of the complete set of N proteins.

2 Train the method on the remaining N–1 proteins (the training set).

3 Predict the secondary structure for the protein taken out.

4 Repeat step 1–3 for all N proteins and assess the average accuracy.

It is possible to test the method by averaging the predictions over all combinations of x proteins ($1<x<N$), each time using the method trained on the remaining $N-x$ proteins. This provides an impression of the influence of different training sets on the sustained accuracy of a single protein being predicted. As the number of combinations grows rapidly with x, the training phase of most methods is too slow for extensive testing using this mode. It can, however, also be used to save computation time if the database is split evenly in test groups of sequences (e.g. 9), as each sequence within a test group is associated with a single training set, thus saving training overhead.

An additional problem in secondary structure prediction is the standard of truth. Most prediction methods are assessed in accuracy by using known tertiary structures from the protein data bank (7) with their secondary structural elements assigned using the DSSP method of Kabsch and Sander (10). Colloc'h *et al.* (11) compared three such secondary structure determination algorithms, among which was the DSSP method, and found significant differences in their secondary structural assignments. This ambiguity in secondary structural assignments can be dramatic for particular proteins where agreement of the methods can be as low as 65% (12, 13). Moreover, in structurally equivalenced sets of homologous proteins with known tertiary structure, the corresponding secondary structural elements can vary in length or show shifts of one to a few residues, and hence a realistic maximum prediction accuracy per residue would be in the range 80–100% (14). Many researchers have suggested that prediction evaluation should be based on the overlap of predicted and observed segments rather than on individual positions (15–20). A recent secondary structure assignment program that combines many of the features of earlier methods, such as checking hydrogen bonding patterns and stereochemical characteristics, is the knowledge-based method STRIDE (21), claimed to yield assignments in close agreement to those made by crystallographic experts.

3 Prediction methods for globular proteins

3.1 The early methods

Attempts to predict protein secondary structure began more than four decades ago (e.g. 22, 23), while the first computer algorithms appeared a quarter of a century ago (24–26). The algorithms of Nagano (22) and Chou and Fasman (25) were based on statistical information, whereas Lim's method (26) was stereochemically oriented and relied on conserved hydrophobic patterns in secondary structures such as amphipathicity in helices (8). Secondary structure prediction has generally been formulated for three states, helix, strand, and coil. This holds also for recent versions of the early and popular GOR method (27, 28), which considers the influence and statistics of flanking residues on the conformational state of a selected amino acid to be predicted. The popular early methods by Lim (26) and Chou–Fasman (25) as well as the GOR method (27, 28) will be described in more detail.

3.1.1 Lim

Lim (26) developed a set of complicated stereochemical prediction rules for α-helices and β-sheets based on their packing as observed in globular proteins. Apart from being the most successful early method (see below), Lim's stereochemical rules are quite important for understanding protein folding. An example is the set of hydrophobicity rules for α-helices with terminal hydrophobic pairs at sequence positions i and $i + 1$, hydrophobic pairs in middle helical segments positioned at $(i, i + 4)$ and middle hydrophobic triplets positioned at $(i, i + 1, i + 4)$ or $(i, i + 3, i + 4)$ (see also *Figure 3*). The Lim method never gained widespread popularity because a computer implementation has not been available until recently.

3.1.2 Chou–Fasman

The most widely used pioneering method is the one by Chou and Fasman (25), in which predictions are based on differences in residue composition for three states of secondary structure: α-helix, β-strand, and turn (i.e. neither α-helix nor β-strand). Chou and Fasman performed a statistical analysis over a number of crystallographically determined protein tertiary structures and determined the frequency of each amino acid type in the three states. The position of turn residues was included in the frequency calculations given significant positional differences in residue type occurrences at turn sites. The frequencies were normalized to amino acid type preferences for each of the structural states by dividing each by that found in all positions of the known structures. For helix and strand, effects of neighbouring residues in the protein sequence were taken into account by averaging the preferences over three residues for α-helix predictions and over two for β-strands. Secondary structures were initiated according to the higher preference values and minimum nucleation lengths required for each structural state. Extensions were effected as long as preferences remained high and certain residues were not encountered (e.g. proline in a α-helix). The Chou–Fasman method has owed its early popularity to the straightforward underlying statistics that are easy to understand.

3.1.3 GOR

The GOR method quickly became the standard for a decade after its first appearance. Although the initial versions GOR I and GOR II predicted four states by discriminating between coil and turn secondary structures, GOR III (28) and the most recent version, GOR IV (29) perform the common three-state prediction. The GOR method relies on the frequencies observed for residues in a 17-residue window (i.e. eight residues N-terminal and eight C-terminal of the central window position) for each of the three structural states. The amino acid frequencies are exploited using an information function based on conditional probabilities defined as:

$$I(S; R) = \log[P(S|R)/P(S)],$$

where S one of the structural states H, E, or C, and R is one of the 20 residue types. The factor $P(S|R)$ denotes the conditional probability of a secondary structural state for a sequence position given that it is occupied by residue type R. Rewriting the formula for frequencies gives:

$$I(S; R) = \log[(f_{S,R}/f_R) / (f_S/N)],$$

where $f_{S,R}$ is the frequency of residue type R in state S, f_R the general frequency of residue type R, and $f_{S/N}$ that of structural state S. Significant in this formula is that the information of a particular residue type in one of the structural states is not only based on the normalized frequency, but shows an extra weighting based on the inverse fraction of all residues in that state. In the GOR method, this formula is used to calculate the information difference between the various states defined as $I(\Delta S; R) = I(S; R) - I(!S; R)$ with !S denoting all other states (*not* S). The information difference formula then becomes:

$$I(S; R) = \log[f_{S,R}/f_S] - \log[f_{!S,R}/f_{!S}].$$

The above formula is defined for a single sequence position, but can be easily extended to the GOR 17-residue window by, for example, writing R_{17} instead of R. Unfortunately, it is not feasible to sample all possible 17-residue fragments directly from the PDB (as there are 17^{20} possibilities). The subsequent versions of the GOR method over the years have explored increasingly detailed approximations of this sampling problem, along with the increase of data in the PDB:

(a) GOR I just treated the 17 positions in the window independently, and so single-position information could be summed over the 17-residue window.

(b) GOR II did the same but sampled over a larger database.

(c) GOR III (28) refined by including pair frequencies derived from 16 pairs between each non-central and the central residue in the 17-long window. As the PDB at the time was not large enough to provide sufficient data, dummy frequencies were calculated (28).

(d) The current version, GOR IV (29) uses pairwise information over all possible paired positions in a window (there are $17 \times 16/2$ possibilities), albeit with a relatively small weight as compared with the GOR I-type single-position information (a) which is included as well.

The theoretical principles used in the GOR method are statistically sound and no *ad-hoc* rules or artificial variables are invoked, which makes it one of the most elegant methods with a high accuracy given its single sequences prediction. However, as in many other methods (*vide infra*), a post-processing step was introduced for the GOR IV method to refine the predictions. Helices are required to be at least four residues in length and strands should consist of two or more residues. If a shorter helix or strand fragment is initially predicted, the method assesses the probabilities of extending the fragment to the minimum associated length or deleting it (i.e. changing it to coil).

3.2 Accuracy of early methods

The Chou–Fasman, GOR III, and Lim methods were assessed to show accuracies of 50%, 53%, and 56% respectively (30). Version IV of the GOR method, however, raises the single sequence prediction accuracy to 64.4% (29), as assessed through jackknife testing (see *Protocol 1*) over a database of 267 proteins with known structure. Random prediction would yield about 40% correctness given the observed distribution of the three states in globular proteins (with roughly 30% helix, 20% strand, and 50% coil). Although they are significantly beyond the random level, these single-sequence prediction accuracies are not sufficient to allow the successful prediction of the protein topology.

3.3 Other computational approaches

The Chou–Fasman and GOR methods both exploit compositional biases exhibited by the three types of secondary structures. Information derived from single sequences have been explored as well in the form of sequence pattern matching (16, 31–34).

On the algorithmic side, researchers have integrated novel computational concepts to optimize the implementation of observed patterns in mapping the primary on to the secondary structure and to thus enhance the success rate of prediction. These include:

(a) Neural network applications (9, 35).

(b) Nearest-neighbour methods (36–39).

(c) Linear discriminant analysis (40).

(d) Inductive logic programming (ILP) (41).

Examples of the first three formalisms will be described in the following section. The latter computational concept (ILP) is designed for learning structural relationships between objects. Muggleton *et al.* (41) used the ILP computer program Golem to automatically describe qualitative rules for residues in the α-helix conformation and central in a 9-residue window. The rules made use of the physico-chemical amino acid characterizations of Taylor (42) and were established during iterative training steps over a small set of 12 known α/α protein structures. With the thus obtained set of rules, α-helices in four independent α/α proteins were predicted with an accuracy of 81% on a per residue basis (Q_3). The Golem algorithm is of limited use because it is only able to predict helices in all-helical proteins.

3.4 Prediction from multiply-aligned sequences

In 1987, Zvelebil *et al.* (43) for the first time exploited multiple alignments to predict secondary structure automatically by extending the GOR method and reported that predictions were improved by 9% compared to single sequence prediction. Also Levin *et al.* (44) quantified the effect and observed 8% increased accuracies when multiple alignments of homologous sequences with sequence identities of $\geq 25\%$ were used. As a result, the current state-of-the-art methods all use input information from multiple sequence alignments.

3.4.1 Neural network methods

Neural networks are organized as interconnected layers of input and output units, and can also contain intermediate (or 'hidden') unit layers (for a review, see ref. 45). Each unit in a layer receives information from one or more other connected units and determines its output signal based on the weights of the input signals. A neural network can be regarded as a black box, which is trained to optimize the grouping of a set of input patterns into a set of output patterns by adjusting the weights of the internal connections. Therefore, neural nets are learning systems based upon complex non-linear statistics.

PHD

The PHD method (**Pro**file network from **HeiD**elberg) (9) combines the added information from multiple sequence information with the optimization strength of the neural network formalism. The method makes use of three consecutive complete neural networks:

(a) The first network produces the first raw 3-state prediction for each alignment position. It takes as input the fractions of the 20 amino acids at each multiple alignment position together with the two 6-residue flanking regions; i.e. a 13-residue window ($w = 13$) is used to predict each alignment position with the central residue in the middle position. The output of the first network for each alignment position is three probabilities for three the states (helix, strand, and coil).

(b) A second network refines the raw predictions of the first level by filtering the 3-state probabilities for each alignment position based on the probabilities of the flanking positions. It takes as input the output of the first network and processes the information using a 17-residue window. The output of the second network comprises for each alignment position the three adjusted state probabilities. This post-processing step for the raw predictions of the first network is aimed at correcting unfeasible predictions and would, for example, change (HHH**EE**HH) into (HHH**HH**HH).

(c) The first two networks perform the basic prediction of the secondary structure associated with a query multiple alignment. However, as the networks can be trained in various ways, PHD employs a number of separately trained consecutive network pairs ((a) and (b)) and feeds their predictions (3-state probabilities) into a third network for a so-called jury decision.

The predictions obtained by the jury network undergo a final filtering to delete predicted helices of one or two residues and changing those into coil. The method was trained on a non-redundant set of 130 alignments from the HSSP database (46), each containing one sequence with a known structure. The method showed an overall prediction accuracy of 70.8% in a jackknife test over 126 alignments (4 of 130 alignments were transmembrane protein families), which for computational reasons were divided in 7 groups (see *Protocol 1*). Although this count is not the highest accuracy reported, the PHD method to date shows the most sustained performance as compared with all other methods available on the Web.

If the PHD webserver is given a single sequence for prediction, it performs a BLAST-search to find a set of homologous sequences and aligns those using the MAXHOM alignment program (46). The resulting alignment is then fed into the actual PHD neural net algorithm.

Pred2ary

Another accurate profile and neural net-based prediction method is Pred2ary (35) which was assessed with an accuracy of 74.8% and balanced prediction over the three structural states. The method employs a second neural net to filter the raw predictions of the first net, as does the PHD method (9). A recent extended version, which combines in a jury decision the outputs of a massive number of 120 networks individually trained, is claimed to predict 75.9% ± 7.9% accurately. This is achieved by constructing *a priori* probabilities of correctly predicting the structural state at each query sequence position for all combinations of network output weighs for helix and strand. These probabilities are then used for a final state prediction corresponding to the highest of the *a priori* probabilities for each of the three states.

3.4.2 k-nearest neighbour methods

As with neural network methods, the application of a k-nearest neighbour method requires an initial training phase in which a large pool of so-called exemplars is established. In the context of secondary structure prediction, this pool typically consists of sequence fragments of a certain length derived from a database of known structures, so that the central residue of such fragments (exemplars) can be assigned the true secondary structural state as a label. Then, a window of the same length is slid over the query sequence and for each window the k most similar fragments from the pool of exemplars are determined using a similarity criterion. The distribution of the k secondary structure labels is then used to derive propensities for the three states. In the methods covered below, k is in the range 25–100.

Yi and Lander

Yi and Lander (36) were the first to use nearest-neighbour classifiers for prediction of secondary structure. A database of 110 proteins with known tertiary structure was used to derive a large collection of 19-residue exemplars for which only the environmental states were noted; i.e. the residue type information was discarded. As a label for each exemplar the secondary structural state of the central residue was taken. For each 19-residue window of the query protein, 50 nearest neighbour exemplars were identified using the amino acid environmental scoring system of Bowie *et al.* (47), which includes as parameters the secondary structure state, accessible surface area and polarity; and scores the likelihood of a residue type to be in a particular state (or range) over these three parameters. As a score, the average was taken of 19 residues within a query window matched with the 19-position exemplar considered. During training, for each exemplar a cut-off score was determined, which should be met by the query fragment compared to it in order to count the exemplar as a neighbour: The cut-off score can be viewed as a reliability check for the predictive value of the exemplars. The 50 thus obtained nearest neighbours showed a distribution of the associated secondary structure labels, from which probability estimates for the three structural states were derived for the query fragment considered. Yi and Lander explored various scoring systems and found that the best performer included 15 environmental classes (3 secondary structures times 5 different accessibility/polarity classes) combined with an amino acid exchange score from the Gonnet *et al.* matrix (48). Note that for this final scoring system, the amino acid types of the exemplars were taken into account. This scenario resulted in a prediction accuracy of 67.1%. Using a neural network for a jury decision over six different scoring systems led to the final accuracy of 68%, as assessed through jackknife testing (*Protocol 1*).

NNSSP

The NNSSP (Nearest Neighbour Secondary Structure Prediction) (37) method adopts the nearest neighbour approach of Yi and Lander (36) for single sequence prediction. Differences with the Yi and Lander method are:

(a) Predictions are made for multiple alignments.

(b) N- and C-terminal positions of helices and strands; and β-turns are explicitly taken as additional secondary structure types.

(c) When predicting, the database of exemplars (see above) is restricted to sequences similar to the query sequence. This reduces computation and leads to more biologically related nearest neighbours,

(d) Alignment regions with insertions/deletions are explicitly taken into account.

Salamov and Solovyev (37) explored various window lengths and finally choose predictors combining window sizes of 11, 17, or 23; nearest neighbour numbers of 50 or 100, and balanced or non-balanced training (i.e. $3 \times 2 \times 2 = 12$ predictors). A simple majority rule over the 12 predictors increased the accuracy by 0.9%. A few simple filters were effected to refine the thus obtained predictions as follows:

(a) Helices predicted to consist of 1 or 2 residues are deleted (changed to coil), but (EHE) becomes (EEE).

(b) Strands of length 1 or 2 are deleted, but (HEEH) becomes (HHHH).

(c) Helices of length 4 or less are deleted. This rule is applied after a full cycle of rule (a) and (b).

The overall accuracy of the method is 72.2%, which results from a jackknife test over the database of 126 proteins by Rost and Sander (9).

PREDATOR

The PREDATOR method of Frishman and Argos (38, 39) owes its accuracy mostly to the incorporation of long-range interactions for β-strand prediction and attains 68% prediction accuracy for single sequence prediction which was assessed using a one-at-a time jackknife test (see *Protocol 1*) over the protein set of Rost and Sander (RS) (9). Using a k-nearest neighbour approach (with $k = 25$ and 13-residue windows), propensities for the general three states (P^H, P^E, and P^C) were determined for each residue. Using pairwise potentials involving long-range interactions, two more propensities for β-strand were determined. This was done by assessing the likelihood for all pairwise 5-residue fragments (separated by more than six amino acids) to form parallel or anti-parallel β-bridges, based on summing residue hydrogen bonding propensities obtained from known structures (two sets of propensities for anti-parallel and one for parallel bridges). As the final parallel and anti-parallel β-strand propensity for each residue (P^{Par} and $P^{Antipar}$), the maximum scoring window pair was taken with the residue considered at the N-terminal position in one of the windows. Pairwise hydrogen bonding potentials were also determined for α-helical residues at a sequence separation of four residues. Their sum was calculated over a 7-residue window to arrive at an extra helix propensity for the residue N-terminal in the window (P^{Helix}). The last additional propensity concerned β-turns (P^{Turn}) and was obtained by summing single-residue propensities in classic β-turn positions 1–4 (49) using a four-residue window. For each of the thus obtained seven independent propensities, threshold values (T) were calculated and used in the following five rules applied consecutively to arrive at a three-state prediction for each residue:

1. If ($P^{Par} > T^{Par}$ or $P^{Antipar} > T^{Antipar}$) and $P^{Helix} < T^{helix}$, then predict β-strand; otherwise, if $P^{Helix} > T^{helix}$, then predict α-helix, otherwise predict coil.

2. If $P^C > T^C$, then predict coil.

3. If $P^E > T^E$, then predict β-strand.

4. If $P^H > T^H$, then predict α-helix.

5. If $P^{Turn} > T^{Turn}$, then predict coil.

Apart from the novel scheme to employ long-range interaction to aid strand prediction, the PREDATOR method can also use information from multiple sequences to enhance predictions. However, PREDATOR does not use or construct a multiple alignment, but rather compares the sequences using pairwise local alignments (50). The current method is not able to extract local alignments from a multiple alignment provided by the user, while leaving the multiple alignment intact, but it is planned to realize this option in a future release (Frishman, personal communication). As predictions by PREDATOR are carried out for a single base sequence, a set of highest scoring local alignments is compiled through matching the base sequence with each of the other sequences. A weight is then compiled for each matched local fragment based on the alignment score and length of the local alignment. For each residue in the base sequence, the weighted sum over all stacked fragments (see *Figure 4*) is compiled independently for the seven propensities and subjected to the above five rules to arrive at a three-state prediction. The extra information conferred by the multiple sequences resulted in a Q_3 of 74.8% (39), as assessed using one-at-a-time jackknife testing over the RS protein set. As for the Pred2ary method showing identical accuracy, this Q_3 is the highest reported in the literature.

3.4.3 Linear discriminant analysis: the DSC method

The DSC method combines the compositional propensities from multiple alignments with a set of concepts important for secondary structure prediction

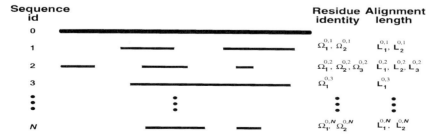

Figure 4 Usage of local alignments in the PREDATOR algorithm. For details, see text.

(see Section 1.3). This information is processed using linear statistics. Apart from the conformational propensities, the following concepts are used:

- N-terminal and C-terminal sequence fragments are normally coil.
- Moments of hydrophobicity (see *Figure 3*).
- Alignment positions comprising gaps are indicative for coil regions.
- Moments of conservation.
- Autocorrelation.
- Residue ratios in the alignment.
- Feedback of predicted secondary structure information.
- Simple filtering.

The relative importance of these concepts was determined in five runs, which successively relied on increased information as follows:

(a) **Run 1**: The GOR method was used on each of the aligned sequences and the average GOR score for each of the three states was compiled for each alignment position.

(b) **Run 2**: For each position in the query multiple alignment, a so-called attribute vector was compiled, consisting of 10 attributes: three averaged GOR scores for H, E, and C; distance to alignment edge; hydrophobic moment assuming helix; hydrophobic moment assuming strand; number of insertions; number of deletions; conservation moment assuming helix and that assuming strand.

(c) **Run 3**: Positional 20-attribute vectors were determined consisting of the above 10 attributes and those in a smoothed fashion.

(d) **Run 4**: Positional 27-attribute vectors were compiled comprising the 20 attributes of the preceding round, combined with fractions of predicted α-helix and β-strand, and fractions of the five most discriminating residue types; His, Glu, Gln, Asp, and Arg.

(e) **Run 5**: A set of 11 filter rules were employed for a final prediction, such as, for example, ([E/C]C**E**[H/E/C][H/C]) → C. These filter rules were found automatically using machine learning.

For run (b) to (d), a linear discrimination function was determined for each of the three secondary structural states. A linear discrimination function is effectively a set of weights for the attributes in the positional vector, so that the secondary structure associated with the highest scoring discrimination function is assigned to the alignment position considered.

The DSC predictions are based on the information arising from the five above runs. The Q_3 was assessed for successively increasing numbers of runs (run 1, runs 1 and 2, runs 1–3, 1–4, 1–5) for the five runs based on the Rost–Sander protein set and comprised 63.5%, 67.8%, 68.3%, 69.4%, and 70.1% (DSC), respectively. The DSC method performs especially well for moderately sized proteins in the range 90–170 residues. A special feature of the DSC technique is that it accepts predictions by the PHD algorithm as input and attempts to refine those using the above concepts. The Q3 of this PHD-DSC combinatorial procedure was evaluated at 72.4% (40).

3.4.4 SSPRED: a secondary structure specific exchange method

The SSPRED method (51) exploits an alternative aspect of the positional information provided by multiple alignment, in that it uses the amino acid pairwise exchanges observed for each multiple-alignment positions. Using the 3D-ALI database (52) of combined structural and sequence alignments of distantly homologous proteins, three amino acid exchange matrices were compiled for helix, strand, and coil, respectively. Each matrix contains preference values for amino acid exchanges associated with its structural state as observed in the 3D-ALI database. They are used to predict the secondary structure of a query alignment through listing the unique observed residue exchanges for each alignment position and adding the corresponding preference values over each of the three exchange matrices. The fact that each exchange type is counted only once for each alignment position provides implicit weighting of the sequences, thus avoiding predominance of related sequences. The secondary structure associated with the matrix showing the highest sum is then assigned to the alignment position. Following these raw predictions, three simple cleaning rules are applied and completed in three successive cycles:

(a) **Single position interruptions**: if a sequence site is predicted in one structural state and the two flanking positions in another, the position is changes into that of the consistent flanking sites, for example (H[E/C]H) becomes (HHH) where [E/C] indicates E or C.

(b) **Double position interruptions**: if in five consecutive positions two middle sites are of another type than the three flanking sites, the middle positions are changed to the flanking types. For instance, (HH[E/C][E/C]H) or (H[E/C][E/C]HH) becomes (HHHHH).

(c) **Short fragments**: helices predicted less than or equal to 4 and strands less than or equal to 2 in length are changed into coil predictions.

The accuracy of the method was assessed over one-at-a-time jackknife testing and amounted to 72%, albeit over a relatively small test set of 38 protein families.

3.5 A consensus approach: JPRED

The JPRED server at the EMBL-European Bioinformatics Institute (Hinxton, UK) conveniently runs state-of-the-art prediction methods such as PHD (9), PREDATOR (38, 39), DSC (40), and NNSSP (37), while also ZPRED (43) and MULPRED (Barton, unpublished) are included. The NNSSP method has to be activated explicitly, as it is the slowest of the ensemble. The server accepts a multiple alignment and predicts the secondary structure of the sequence on top of the alignment: Alignment positions showing a gap for the top sequence are deleted. A single sequence can also be given to the server. In the latter case, a BLAST-search is performed to

CLUSTALX:

```
3chy          ADKELKFLVVDDFSTMRRIVRNLLKELGFNNVEEAEDGVDALNKLQAGSYGVFVISDWNMPNMDGLELLKTIRADGAMSALPVLMVTAEAKKENIIAAAQAGASGYVVKPFTAATLEEKLNKIFEKLGM
FLAV_MEGEL    MVEIVYWSGTGNTEAMANEIEAAVKAAGVESVRFEDTNVDDVALLGCPAMGSEELEDSVVEPFFTDLAPKLKGKSYGWGSGEWMDAWKQRTEDFGATVIGTAIVNEMPNAPECKE.LGEAAAKA..
4fxn          .MKIVYWSGTGNTEKMAELIAKGIIESGVNTINVSDVNIDELLILGCSAMGDEVLEESEFEPFIEEISTKISGKGSYGWGDGKWMRDFFERMNGYGCVVVETPLIVNEPLEAEQDCIEFGKKIANI..
FLAV_DESGI    KALIVYGSTTGNTEGVAEAIAKTLNSEGTTVNVADVTAPGLALLGCSTWGDSEDLQEDFVPLYEDLDRAGLKDKGCGDSSYTYAVDVIEKKAEELGATLVASSLKIGEPLSAE..VLDWAREVLARV
FLAV_DESSA    KSLIVFGSTTGNTETAAEYVAEAFENKEVELKNVTDVSVADLGFLGCSTWGEEELQDDFIPLYDSLENADLKGKGCGDSDSYTYAVDAIEELEKMGAVVIGDSLKIGDPERDE..IVSWGSGIADKI
FLAV_DESDE    KVLIVFGSSTGNTESIAQKLEELIAAGGVTLLNAADASAENLALFGCSAWGMEEMQDDFLSLFEENFRFGLAGRAGDQEYEHAVPAIEERAKELGATIIAEGLKNGAFVDGLKQL
1fx1          KALIVYGSTTGNTEYTAETIARQLANAGVDSRDAASVEAGGLFLLGCSTWGDDEFLQDDFIPLFDSLEETGAQGRGCGDSSYEYAVDAIEEKLKNLGAEIVQDGLRIGDPLAARDDIVGWAHDVRGAI
FLAV_DESVH    KALIVYGSTTGNTEYTAETIARELADAGYDVSRDAASVEAGGLFLLGCSTWGDDEFLQDDFIPLFDSLEETGAQGRGCGDSSYEYAVDAIEEKLKNLGAEIVQDGLRIGDPLAARDDIVGWAHDVRGAI
FLAV_CLOAB    KISILYSSKTGKTERVAKLIEEGVKRSGVKTMNLDAVDKFLQIFGTPTYAN.ISWEMKKWIDESSEFNLEGKSTANSIAGGALLTILNHLMVKGMLVYSGGVHIEQNEDENARIFGERIANKVK
FLAV_ANASP    KIGLFYGTGKTESVAEIIRDEFGNDV..LHDVSQAEVTDLNIIGCPTWNICQ...WEGLYSELDDVDFNGKGTGDQIGYAAIGIILEEKISQRGKTVGYWGKFGLALDEDNQSDLTDDRIKSWVA
FLAV_AZOVI    KIGLFFGSNTGKTRKVAKSIKKRFDDET..ALNVNRVSAEDFAILGTPTLGEGPNESWEEFLPKIEGLDFSGKGLGDVGYGPALGELYSFFKDRGAKIVGSWKFGLALDLDNQSGKTDERVAAWLA
2fcr          KIGIFFSTSTGNTTEVADFIGKTLGAKA..IDVDDVTDPQALKFLGAPTWNTGT...SWDEFLYKLPEVDMKDLGLGDAEGYPAIEEIHDCFAKQGAKPVGFSGKFGLPLDMVNDQIPMEKRVAGWVE
FLAV_ENTAG    TIGIFFGSDTGQTRKVAKLIHQKLDGIA..LDVRRATREQFLSLLGTPTLGDGPQYDSWQEFTNTLSEADLTGKGLGDLNYSAMRILYDLVIARGACVVGNWNEFGLPLDQENQYDLTEERIDSWLE
FLAV_ECOLI    ITGIFFGSDTGNTENIAKMIQKQLGKDV..VHDIAKSSKEDLELLGIPTWYGQ....WDDFFPTLEEIDFNGKGCGDQEDYAALGTIRDIIEPGATIVGHWDHFGLA.DEDRQPELTAERVEKNVK
dsc           --EEEE----------------HHHHHHHHHHHH-------------------------------------------HHHHHHHH-----------------------------------HHHHHHH---
mul           -EEEEE----------HHHHHHHHHHHH----------------------------------------------HHH-HHHHHHHH-----------------HEE------------------HHHHHHHHHHHHH----
phd           --EEEEE------------HHHHHHHHHHHHH--------------------------------------------HHHHHHHHHHH----EEE----------------------------HHHHHHHHHHHH-
pred          EEEEEE----------HHHHHHHHHHHH----------------------------------------------HHHHHHHHHHHH-------EEEE-------------HHHHHHHHHHHHH-
zpred         EEEEEE----------HHHHHHHHHHHHH---------------------------------------------HHHHHHHHHHHH------EEEE-----------------HHHHHHHHHHHH---
```

PRALINE:

```
3chy          ADKELKFLVVDDFSTMRRIVRNLLKELGFNNVEEAEDGVDALNKLQAGSYGVFVISDWNMPNMDGLELLKTIRADGAMSALPVLMVTAEAKKENIIAAAQAGASGYVVKPFTAATLEENLNKIFEKLGM
1fx1          ..PKALIVYGSTTGNTEYTAETIARQLANAGVDSRDAASVEAGGLFEGFDLVLSTWGDDS.DFIPLFDSLEETGAQGRKVACFGCGDS.SEYFCGAKNLGABGLRIDRLARDDIVGWAHDVRGAI.
FLAV_DESDE    .MSKVLIVFGSSTGNTESIAQKLEELIAAGGHETLLNAADASAENLADGYDAVLSAWGMED.DFLSLFEENFRFGLAGRKVAAFASGDQ.SEHFCGAKNLGABGLRIDRLARDDIVGWAHDVRGAI.
FLAV_DESVH    .MPKALIVYGSTTGNTEYTAETIARELADAGYDVSRDAASVEAGGLFEGFDLVLSTWGDDS.DFIPLFDSLEETGAQGRKVACFGCGDS.STYFCGAEELGATSLKIDD.SAEVLDWAREVLARV.
FLAV_DESGI    .MPKALIVYGSTTGNTEGVAEAIAKTLNSEGMTVNVADVTAPGLALFEGFDLVLSTWGDDS.DFIPLYDSLENADLKGKGCGDS.DTYFCGAEKMGAVSLKIDE.RDEIVSWGSGIADKI.
FLAV_DESSA    .MSKSLIVYGSTTGNTETAAEYVAEAFENKEIDELKNVTDVSVADLGNGYDIVLSTWGEEE.DFIPLYDSLENADLKGKKVSVFGCGDS.DTYFCGAEKMGAVSLKIDE.RDEIVSWGSGIADKI.
4fxn          .MK..IVYWSGTGNTEKMAELIAKGIIESGKDNTINVSDVNIDELLNE.DILIISAMGDEV.EFEPFIEEIS.TKISGKKVALFG.....SGWGDGKWMDAWKQRTEDFGAT.AIVNEMPNAPECKE
FLAV_MEGEL    .MVE..IVYWSGTGNTEAMANEIEAAVKAAGADESVRFEDTNVDDVASK.DVILPAMGSEE.VVEPFFTDLA.PKLKGKKVGLFG.....SGWGSGEEDTGAT.AIVNDEA.PECKELGEAAAKA.
2fcr          ..KIGIFFSTSTGNTTEVADFIGKTLGAK.ADPIDVD.DVTDPQALKDYDLLFPTWNTG.WD.EFLYDKLPEVDMKDLGVAIFGLGDAEGDNFCDAAKQGAKGLPLDIFMEKRVAGWVEAVVSETGV
FLAV_ANASP    .SKKIGLFYGTGKTESVAEIIRDEFGND.VVLHDVS.QAE.VTDLNDYQYLIPTWNIGEWE.GLY.SELDDVDFNGKLVAYFGTGDQIGDNFQDASQRGGRKGLALDD.TDDRIKSWVAQLKSEFGL
FLAV_ECOLI    ..AITGIFFGSDTGNTENIAKMIQKQLGKD.VAVHDIA.KSS.KEDLEAYDILLFPTWYGEWD.DFF.PTLEEIDFNGKLVALFGCGDQEDEYFCDAEPRGATGLAIDE.TAERVEKWVKQISEELHL
FLAV_AZOVI    .AKIGLFFGSNTGKTRKVAKSIKKRFDDETMSALNVN.RVS.AEDFAQYGFLIPTLGEGEWE.EFL.PKIEGLDFSGKTVALFGLGDVGENYLDAKDRGAKGLALDGT.DERVAAWLAQIAPEFGL
FLAV_ENTAG    .MATIGIFFGSDTGQTRKVAKLIHQKLDGI.ADPLDVR.RAT.REQFLSYPVLLPTLGDGEWQ.EFT.NTLSEADLTGKTVALFGLGDQLNKNFVSAIARGACGLPLDD.TEERIDSWLEKLKPAV.L
FLAV_CLOAB    .MKISILYSSKTGKTERVAKLIEEGVKRSGNIVKTMNLDAVDKKFLQESEGIIPTYYAN.WEMKKWIDESSEFNLEGKLGAAFSTANSIASDI.AMVKGMLEIQDNR.FGERIANKVKQIF.
dsc           --EEEEEE------------HHHHHHHHHHHH-------HHHHH-------HHHEH-----EEEE----------HHHHHHHHHHH--------------HHHHHHHHH-H----
mul           --EEEEEE----------HHHHHHHHHHHH---------HHH-----H--HH-HHHHH-----HEEE--------HHHHHHHHHH------EE----E------HHHHHHHHHHHHH------
phd           --EEEEEE---------HHHHHHHHHHHH----------HHH-----EE-EE------HHH-----HHHHH--------HHHHHHHH------E-----HHHHHHHHHHHHH-
pred          --EEEEE-----------HHHHHHHHHHH----------HHHHH-------EEE-----HHHH-------------EEEE-----------HHHHHHHHHHHH-------EEE------EEE-HHHHHHHHHHHH-
zpred         --EEEEE-----------HHHHHHHHHHH----------HHHHH-------EEE-----HHHHHH-----------EEEE-----------HHHHHHHHHHHH-------EEE------EEE-HHHHHHHHHHHH-
```

JPRED CONSENSUS PREDICTION:

3chy		ADKELKFLVVDDFSTMRRIVRNLLKELGFNNVEEAEDGVDALNKLQAGSYGVFVISDWNMPNMDGLELLKTIRADGAMSALPVLMVTAEAKKENIIAAAQAGASGYVVKPFTAATLEEKLNKIFEKLGM
cons	CLUSTALX 53.9%	-EEEEEE-----HHHHHHHHHHHHH----
cons	PRALINE 85.1%	---EEEEEE-----HHHHHHHHHH----
cons	HOMOLOGS 88.3%	---EEEEE----EEEHH---HHHHHHHHHH----
DSSP		EEEE HHHHHHHHHHHH

Figure 5 Secondary structure prediction for chemotaxis protein cheY (3chy). The top alignment block represents the multiple alignment of the 3chy sequence with 13 distant flavodoxin sequences by the method PRALINE. The middle block is the same sequence set aligned by CLUSTALX. Under each of the alignments are given the alignments by five secondary structure prediction methods. The bottom block depicts consensus secondary structures determined by Jpred over the five methods used, respectively for the PRALINE and CLUSTALX alignments, as well as for a set of 32 homologous sequences aligned by CLUSTALX (cons HOMOLOGS). Vertical bars (|') under each of the consensus predictions indicate correct predictions. The bottom line identifies the standard of truth as obtained from the 3chy tertiary structure by the DSSP program. (10) The secondary structure states assigned by DSSP other than 'H' and 'E' were set to ' '(coil) for clarity.

find homologous sequences, which are subsequently multiply aligned using CLUSTALX and then processed with the user-provided single sequence on top in the alignment. If sufficient methods predict an identical secondary structure for a given alignment position, that structure is taken as the consensus prediction for the position. In case no sufficient agreement would be reached, the PHD prediction is taken. This consensus prediction is somewhat less accurate when the NNSSP method is not invoked or completed in the computer time slot allocated to the user. An example of output by the JPRED server for the signal transduction protein cheY (PDB code 3chy) is given in *Figure 5* (*vide infra*).

3.6 Multiple-alignment quality and secondary-structure prediction

Multiple-alignment protocols use heuristics to overcome the combinatorial explosion that arises when all possible alignments would be tested exhaustively. Most global alignment methods therefore establish an order in which the sequences are aligned progressively based on the alignment scores of all possible pairwise alignments (the number is $N \times (N - 1)/2$ with N the number of sequences). Although most methods show a comparable overall quality in alignment construction for sequences showing residue identities of 30% or higher, significant differences can arise in individual cases, particularly when evolutionary distant sequences are included. As the currently most successful secondary structure prediction methods all employ positional information from multiple alignments, it is clear that alignment quality is crucial for accurate prediction. As an example, the popular multiple alignment program CLUSTALX and the recently developed method PRALINE (53) (see below) were used to automatically construct a multiple alignment for the signal transduction protein cheY (PDB code 3chy) and 13 distant flavodoxin sequences. The 3chy structure adopts a flavodoxin fold (see *Figure 2*) despite very low sequence similarities with genuine flavodoxins. *Figure 5* shows both alignments with secondary structure predictions by the JPred server as well as JPred consensus predictions for the two alignments. The difference in accuracy of the two consensus predictions amounts to more than 30%, an order of magnitude more than the increase in prediction accuracy obtained over the last five years. It must be stressed that the flavodoxin sequences are evolutionary distant from the cheY sequence, such that the alignments were only included to illustrate their crucial importance for secondary structure prediction rather than to argue in favour of any of the two used alignment programs based on this single example. The Jpred server was also given the single 3chy sequence, after which the Jpred server constructed an evolutionary related set of homologs through a BLAST search and aligned the 32 resulting sequences using CLUSTALX. The accuracy of the consensus secondary structure prediction by Jpred was 3% higher than that obtained for the PRALINE alignment of the cheY-flavodoxin set (*Figure 5*). Moreover, it successfully delineated the second β strand of the 3chy structure, which was missed by the predictions based on both the CLUSTALX and PRALINE alignments.

3.7 Iterated multiple-alignment and secondary structure prediction

As mentioned, most reliable secondary structure prediction methods utilize sequence information in multiple alignments and their prediction accuracy is crucially dependent on the quality of a multiple alignment used. If in turn a multiple alignment would be guided by the predicted secondary structure, an iterative scheme would be possible that optimizes both the multiple-alignment quality and secondary structure prediction. This procedure is implemented in the PRALINE multiple alignment method (53). A multiple alignment is constructed initially without information about the secondary structure (*Figure 6a*). Then, the secondary structure is predicted (for which any of the aforementioned methods could be used) and iteratively a new alignment is constructed, now using the predicted secondary structure. PRALINE employs dynamic programming to progressively construct a multiple alignment for a query set of sequences and therefore relies on an amino acid exchange weights matrix and a pair of gap penalties (for a review, see ref. 54). The initial alignment is constructed using a default residue exchange matrix (e.g. the BLOSUM62 matrix) and gap penalties. After secondary structure prediction, resulting in a tentative secondary structure for each sequence if a single sequence-based method is used or in a single secondary structure if a method reliant on a multiple alignment is effected (*Figure 6a*), PRALINE utilizes the thus obtained secondary structure information as illustrated in *Figure 6b*. At each alignment step during the progressive alignment, pairs of sequences (and/or profiles representing already aligned sequence blocks) are matched using three secondary structure-specific residue exchange matrices (55) and associated gap penalties. As shown in *Figure 6b*, the residue exchange weights for matched sequence positions with identical secondary structure states is taken from the corresponding residue exchange matrix. Sequence positions with inconsistent secondary structure states are treated with the default exchange matrix. The secondary structure information is thus used in a conservative manner based upon the assumption that consistent secondary structure predictions are indicative for their reliability when performed for each individual sequence (*Figure 6a*). In this way, the multiple alignments guide the secondary structure predictions, which in turn guide the alignment.

(a)

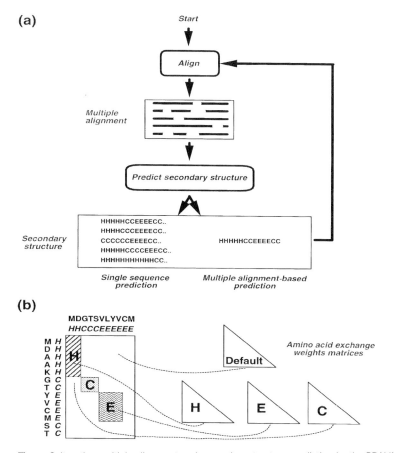

(b)

Figure 6 Iterative multiple-alignment and secondary-structure prediction by the PRALINE method.

4 Prediction of transmembrane segments

Membrane proteins (MP) form a distinct topological class due to the presence of one or more transmembrane (TM) sequence segments. In contrast to globular proteins where all possible mutual orientations of individual structural elements are in principle possible, MP transmembrane segments are subjected to severe restrictions imposed by the lipid bilayer of the cell membrane.

There is a considerable lag in structures available for MPs relative to the large and vastly growing numbers of soluble proteins, as little X-ray or NMR data regarding the tertiary structure of MPs have been available until recently (56). The most frequently observed secondary structure in transmembrane segments is the α-helix, but also transmembrane structures based on β-strands that constitute a β-barrel have been encountered. The initial idea that TM segments are either completely α-helical or consist of β-strands exclusively, was disrupted by electron microscopy data for the nicotinic acetylcholine receptor (57), which was interpreted as a central five-helix bundle surrounded by β-strands, albeit based on preliminary data with low resolution.

133

Fortunately, the location of the transmembrane segments in primary structure of the MP is relatively easy to predict due to the rather strong tendency of certain hydrophobic amino acid types with their special physico-chemical properties to occur in membrane spanning regions. Thus, efforts concerning the theoretical analysis of MPs over the past two decades have been focused on the determination of the membrane sequence segment boundaries and their tentative orientation with respect to the membrane, although mostly assuming α-helical structures.

4.1 Prediction of α-helical TM segments

The following considerations form the basis of transmembrane subsequence prediction.

(a) Amino acids immersed in the lipid phase are likely to be hydrophobic. Therefore, any physical measure of amino acid hydrophobicity derived from physical calculations and/or experimental data can serve as a measure of likeliness for a residue type to occur in a membrane-spanning segment.

(b) The propensities of amino acids to reside in the lipid bilayer can be inferred from abundant but not very precise experimental data on the boundaries of the transmembrane segments acquired from site-directed mutagenesis, enzymatic cleavage, immunological methods and the like. This contrasts with the standard secondary structure prediction methods for soluble proteins where statistical propensities of different amino acids to form one of the major secondary structure elements are derived from more accurate protein tertiary structural data from X-ray crystallography and NMR spectroscopy.

(c) Transmembrane segments are believed to adopt in most cases α-helical conformation. An α-helix is the most suitable local arrangement because, in the absence of water molecules inside the membrane, all main-chain polypeptide donors and acceptors must mutually satisfy each other through formation of hydrogen bonds as occur in an α-helix. This energetic argument is supported by experimental evidence where polypeptide chain tends to adopt helical conformation in a non-polar medium (58). Therefore, α-helical propensities of amino acids derived from the analysis of globular proteins can be considered in MP structure prediction.

Although the globular protein interior is less apolar than the lipid bilayer, extensive usage of these data has been made for MP structure prediction, particularly with the classical hydrophobicity scale of Kyte and Doolittle (59). Other techniques are more specifically aimed at searching MP transmembrane regions (60–63). Hydrophobic scales can be used to build a smoothed curve, often called a hydropathic profile, by averaging over a sliding window of given length along the protein sequence to predict transmembrane regions. Stretches of hydrophobic amino acids likely to reside in the lipid bilayer appear as peaks with lengths corresponding to that expected for a transmembrane segment, typically 16–25 residues. The choice of window length should correspond to the expected length of a transmembrane segment. Given that the average membrane thickness is about 30 Å, approximately 20 residues form a helix reaching from one lipid bilayer surface to another. A further threshold is also required to determine the exact boundaries of a membrane spanning segment. Kyte and Doolitle (59) early on set their limit by examining the hydropathic character of just a few membrane proteins. Later, a much larger learning set was used by Klein *et al.* (64) through discriminant analysis. Rao and Argos (65) suggested a minimum value for the peak hydrophobicity and a cut-off value at either end of the peak to terminate the helix.

Although many of the above techniques constitute an essential part of all major sequence analysis packages, the relatively simple physical considerations forming the basis of these methods do not exhaust the whole variety of possible situations.

If a membrane protein has more than one transmembrane helix, the relative orientation of the helices and the interaction of the corresponding sidechains are also important for structure prediction. The structure of the membrane proteins determined to date and also some theoretical evidence (66) support the view that α-helices in membranes form compact clusters. While the residues facing the lipid environment conform to the preferences described above, the interface residues between different helices do not necessarily have contact with the membrane, and, can therefore, behave differently. It is possible that charged residues occur in the helices in a coordinated fashion such that positively charged sidegroups on one helix will have their negatively charged counterparts on another helix. These charges could, for instance, constitute a membrane channel. Intra-membrane α-helices can thus have an amphipathic character (see above). In such cases hydropathic profiles can work poorly in detecting transmembrane segments. In certain cases where the number of transmembrane segments is large (more than 20 as in some channel proteins), the inner helices of the transmembrane helical bundle can completely avoid contact with the lipid bilayer and, therefore, any restrictions on their amino acid content—or even length—could be artificial.

Eisenberg *et al.* (67) introduced a quantitative measure of helix amphipathicity called the 'hydrophobic moment', and defined as a vector sum of the individual amino acid hydrophobicities radially directed from the helix axis. The hydrophobic moment provides sufficient sensitivity to distinguish amphipathic α-helices of globular, surface, and membrane proteins. Many methods for amphipathic analysis were developed based on Fourier analysis of the residue hydrophobicities (68, 69) and the average hydrophobicity on one helix face (70).

Several prediction methods have emerged which utilize multiple factors, complex decision rules, and large learning sets. Von Heijne (63) proposed a synthetic technique in which a standard hydrophobicity analysis is supplemented by charge bias analysis (see Section 4.2). Other methods include the joint usage of several selected hydrophobicity scales (71) or the application of optimization techniques with membrane segments as defined by X-ray analysis serving as reference examples (72).

Persson and Argos (73) incorporated sequence information from multiple alignments to aid TM prediction. The propensities of amino acids to reside in either the central or the flanking regions of a transmembrane segment were calculated using more then 7500 individual helical assignments contained in the SWISS-PROT sequence databank. These values were then used to build a prediction algorithm wherein, for each segment of a multiple sequence alignment, and for each sequence, average values of the central and flanking propensities are calculated over windows of respectively 15 and 4 residues long. If the peak values for central transmembrane regions exceed a certain threshold, this region is considered as a possible candidate to be membrane spanning. The algorithm then attempts to expand this region in both sequence directions until a flanking peak is reached or the central propensity averages fall below a certain value. Additional restraints are imposed on the possible length of the tentative transmembrane segments. The optimal window length was found to be about 15 residues. Due to the increased amount of information utilized by the technique, more accurate prediction results were achieved as compared with earlier methods. The gain in sensitivity is due to the usage of multiple alignments as well as the introduction of a second propensity for flanking regions.

Neural networks (see Section 3.4.1) have also been applied to the TM prediction problem. Early attempts involved training on secondary structural elements of globular proteins (74). Rost *et al.* (75) trained the PHD method on multiple alignments for 69 protein families with known transmembrane helices and achieved 95% prediction accuracy using the jackknife test.

4.2 Orientation of transmembrane helices

Another aspect of transmembrane segment prediction is prediction of membrane sidedness, or orientation. For bacterial membrane proteins it was found that intracellular loops in between transmembrane helices contain arginine and lysine residues much more frequently than the extracellular exposed loops (76, 77). This pattern has been shown to apply also to eukaryotic membrane proteins, but to a lesser extent (78). An additional observation, made for eukaryotic proteins, is that the difference in the total charge of the approximately 15 residues flanking the transmembrane region on both sides of the membrane also coincides with the orientation of the protein (79). If the C-terminal portion of the protein adjacent to this segment is more positive in charge then the N-terminal portion, the C-terminus will reside in the cytosol, and *vice versa*. Non-random charge distribution may also play an important role in membrane insertion of the protein. These findings, collectively known as the 'positive inside rule', aid prediction schemes for MP topology. However, the positive inside rule is only applicable to α-helical TM regions.

4.3 Prediction of β-strand transmembrane regions

As the methods described above all predict TM segments assuming the α-helical conformation, transmembrane segments constituted by β-strands are not likely to be predicted successfully. Four different families of β-barrel membrane proteins are known to date (porins, OmpA, FhuA, and FepA). For example, porins form voltage-dependent membrane channels and have a β-barrel fold constituted by 16 β-strands (80). Hydrogen bonds are formed only between adjacent β-strands. Most of the outer surface of the barrel faces the lipid environment whereas the internal part serves as an aqueous pore. Each individual β-strand could therefore be expected to be amphipathic with a period of two residues. However, while every second residue facing the lipid bilayer is hydrophobic, those side chains protruding towards the interior of the barrel display no definitive tendency, thus lowering the amphipathic signal. Another complication is the fact that the number of amino acid residues in extended conformation needed to span the membrane is much smaller than that for the helical conformation, typically about 10. Consequently, smoothed hydropathic profiles are likely to miss such short stretches.

5 Coiled-coil structures

If a protein is predicted to contain α-helices, higher-order information as well as increased confidence in predictions made could be gained from testing the possibility that a pair of helices adopts a superhelical twist resulting in a coiled-coil conformation. The left-handed coiled-coil interaction involves a repeated motif of seven helical residues (*abcdefg*). The *a* and *d* positions are normally occupied by non-polar residues constituting the hydrophobic core of the helix–helix interface, whereas the other positions display a high likelihood to comprise hydrophilic residues. The *e* and *g* positions in addition are often charged and can form salt-bridges to each other. The program COILS2 (81, 82) exploits this information and compares a query sequence with a database of known parallel two-stranded coiled-coils. A similarity score is derived and compared to two score distributions, one for globular proteins (without coiled-coils) and one for known coiled-coil structures, and a probability is then calculated for the query sequence to adopt a coiled-coil conformation. As the program assumes the presence of heptad repeats, the probabilities are derived using windows of 14, 21, and 28 amino acids. However, the program offers the option to include user-defined window lengths two allow the handling of cases with extreme coiled-coil lengths. A recently updated scoring matrix which includes new structures with known coiled-coils and contains amino acid type propensities at the various positions in the heptad repeats, led to increased recognition of coiled-coils elements. The COILS2 method accurately recognises left-handed two-stranded coiled coils but loses sensitivity for coiled-coil structures composed of more than two strands. It is not able to recognize right-handed or buried coiled–coil helices and therefore is not applicable to transmembrane coiled-coil structures known to basically show the similar coiled-coil conformations as soluble proteins, albeit with dramatically different and more hydrophobic constituent amino acids (56).

6 Threading

If a homologous protein with known structure is available for a query sequence, this structure can then be aligned to the query sequence using the threading technique (83). Treading methods test the feasibility for a given sequences to adopt a particular fold, based on assessing the likelihood for the amino acids in the query sequence to occur in the local residue environments within the known tertiary structure. The optimal fit of the query sequence through the tertiary structure effectively leads to an alignment, which can be used to copy the secondary structure of the known fold to the query sequence. Although the incorporation of tertiary structure information should lead to better alignment and recognition of related sequences, the increased sensitivity of available threading methods as compared with conventional sequence alignments is not always clear. Jones *et al.* (84) discuss various threading methods available and also how their results should be interpreted.

7 Recommendations and conclusions

Table 1 lists WWW addresses of some of the available prediction methods discussed in this chapter and in *Protocol 2* some recommendations are given to maximize the chances of an accurate prediction of the secondary structure associated with a protein query sequence. In cases where a multiple alignment is used, it is generally important to test the consistency and quality of the alignment constructed, as this can have dramatic consequences for the prediction accuracy of multiple-alignment-based methods. In testing the consistency of the currently most accurate prediction methods and determining a consensus prediction, the positional reliability indices offered by some of prediction methods should be included. Furthermore, the general accuracies for predicting each of the three secondary structural states that are published for a number of the methods can be used to weight the contribution of their positional predictions

Table 1 Websites of various secondary structure prediction methods and related services

Service	Reference	URL
GOR4	27, 28, 29	http://absalpha.dcrt.nih.gov:8008/gor.html
PHD[a]	9	http://dodo.cpmc.columbia.edu/predictprotein/[b]
Pred2ary	35	http://yuri.harvard.edu/~jmc/2ary.html
NNSSP[a]	37	http://dot.imgen.bcm.tmc.edu:9331/pssprediction/pssp.html
PREDATOR[a]	38, 39	http://www.embl-heidelberg.de/cgi/predator_serv.pl
DSC[a]	40	http://bonsai.lif.icnet.uk/bmm/dsc/dsc_read_align.html
Zpred[a]	43	http://kestrel.ludwig.ucl.ac.uk/zpred.html
Jpred	-	http://barton.ebi. ac.uk/servers/jpred.html
COILS2	81, 82	http://www.isrec.isb-sib.ch/coils/COILS_doc.html

[a] Method can also be run using the Jpred server.

[b] Mirror websites for PHD can be found here as well.

in a consensus. Specifically, the PREDATOR method should be included in the trials as it is the only method relying on multiple local rather than global alignment of the query sequences. It is important to realize that there is no single best prediction method so that the degree of consistency over a variety of methods is crucial for getting an idea about the prediction accuracy. Attempts to recognize higher-order structure, such as the fold the protein might adopt or the likelihood for coiled-coil structures, could enhance the confidence in predictions made or help correcting possible mispredictions. Easily recognizable errors might be disruptions in alternating α-helix/β-strand predictions in a likely α/β protein fold or the occurrence of a single β-strand within a tentative α-helical protein. In general, it is likely that the accuracy of computerized prediction methods can be enhanced further if such reasoning with higher order structure is formalized and incorporated in the prediction mechanisms. Some easy benefits will come from the steadily increasing structural protein data that can be used to better train and tune the statistical methods. The current availability of the prediction methods optimizes the chance for development of sensitive consensus methods. It is clear from the ongoing interest and activity in both the application and development of secondary structure prediction methods that the end of the three decades of research efforts is not in sight.

Protocol 2

Predicting secondary structure

1   Get a balanced and non-redundant set of homologous sequences for a given protein query sequence.

2   Try a number of multiple alignment routines to obtain a consistent multiple alignment.

3   Check the alignment carefully by eye using any additional information (e.g. active site residues, disulfide bridges, etc.).

4   Use as many good secondary structure prediction methods as possible and construct a consensus prediction (a convenient aid is the Jpred server).

5   Try to recognize super-secondary or higher-order structural features from the predicted secondary structure elements and try to interpret and correct prediction results (e.g. the missed second β-strand in the flavodoxin example) (see Section 3.6).

References

1. Heringa, J. and Taylor, W. R. (1997). *Curr. Opin. Struct. Biol.*, **7**, 416.
2. Springer, T. A. (1997). *Proc. Natl. Acad. Sci. USA*, **94**, 65.
3. Baldwin R. L. and Roder H. (1991). *Curr. Biol.*, **1**, 218.
4. Goldenberg, D. P., Frieden, R. W., Haack, J. A., and Morrison, T. B. (1989). *Nature*, **338**, 127.
5. Baldwin, R. L. (1990). *Nature*, **346**, 409.

6. Flores, T. P., Moss, D. S., and Thornton, J. M. (1994). *Protein Eng.*, **7**, 31.

7. Bernstein, F. C., Koetzle, T. F., Williams, G. J., Meyer, E. F., Brice, M. D., Rodgers, J. R., *et al.* (1977). *J. Mol. Biol.*, **112**, 535.

8. Schiffer, M. and Edmundson, A. B. (1967). *Biophys. J.*, **7**, 121.

9. Rost, B. and Sander, C. (1993). *J. Mol. Biol.*, **232**, 584.

10. Kabsch, W. and Sander, C. (1983). *Biopolymers*, **22**, 2577.

11. Colloc'h, N., Etchebest, C., Thoreau, E., Henrissat, B., and Mornon, J.-P. (1993). *Protein Eng.*, **6**, 377.

12. Sklenar, H., Etchebest, C., and Lavery, R. (1989). *Proteins: Struct. Funct. Genet.*, **6**, 46.

13. Woodcock, S., Mornon, J.-P., and Henrissat, B. (1992). *Protein Eng.*, **5**, 629.

14. Russell, R. B. and Barton, G. J. (1993). *J. Mol. Biol.*, **234**, 951.

15. Taylor, W. R. (1984). *J. Mol. Biol.*, **173**, 512.

16. Cohen, F. E., Abarbanel, R. M., Kuntz, I. D., and Fletterick, R. J. (1986). *Biochemistry*, **25**, 266.

17. Cohen, F. E. and Kuntz, I. D. (1989). In *Prediction of protein structure and the principles of protein conformation* (ed. G. D. Fasman), pp. 647–706. Plenum, New York, London.

18. Sternberg, M. J. E. (1992). *Curr. Opin. Struct. Biol.*, **2**, 237.

19. Benner, S. A., Cohen, M. A., and Gerloff, D. (1993). *J. Mol. Biol.*, **229**, 295.

20. Rost, B., Sander, C., and Schneider, R. (1994). *J. Mol. Biol.*, **235**, 13.

21. Frishman, D. and Argos, P. (1995). *Proteins: Struct. Funct. Genet.*, **25**, 633.

22. Szent-Györgyi, A. G. and Cohen, C. (1957). *Science*, **126**, 697.

23. Periti, P. F., Quagliarotti, G., and Liquori, A. M. (1967). *J. Mol. Biol.*, **24**, 313.

24. Nagano, K. (1973). *J. Mol. Biol.*, **75**, 401.

25. Chou, P. Y. and Fasman, G. D. (1974). *Biochemistry*, **13**, 211.

26. Lim, V. I. (1974). *J. Mol. Biol.*, **88**, 857.

27. Garnier, J., Osguthorpe, D. J., and Robson, B. (1978). *J. Mol. Biol.*, **120**, 97.

28. Gibrat, J.-F., Garnier, J., and Robson, B. (1987). *J. Mol. Biol.*, **198**, 425.

29. Garnier, J. G., Gibrat, J.-F., and Robson, B. (1996). In *Methods in enzymology* (ed. R. F. Doolittle), Vol. 266, pp. 540-53. Academic Press.

30. Schultz, G. A. (1988). *Annu. Rev. Biophys. Chem.*, **17**, 1.

31. Cohen, F. E., Abarbanel, R. M., Kuntz, I. D., and Fletterick, R. J. (1983). *Biochemistry*, **25**, 4894.

32. Taylor, W. R. and Thornton, J. M. (1983). *Nature*, **354**, 105.

33. Rooman, M. J., Wodak, S., and Thornton, J. M. (1989). *Protein Eng.*, **3**, 23.

34. Presnell, S. R., Cohen, B. I., and Cohen, F. E. (1992). *Biochemistry*, **31**, 983.

35. Chandonia, J.-M. and Karplus, M. (1998). *Proteins: Struct. Funct. Genet.*, **35**, 293.

36. Yi, T.-M. and Lander, E. S. (1993). *J. Mol. Biol.*, **232**, 1117.

37. Salamov, A. A. and Solovyev, V. V. (1995). *J. Mol. Biol.*, **247**, 11.

38. Frishman, D. and Argos, P. (1996). *Protein Eng.*, **9**, 133.

39. Frishman, D. and Argos, P. (1997). *Proteins: Struct. Funct. Genet.*, **27**, 329.

40. King, R. D. and Sternberg, M. J. E. (1996). *Protein Sci.*, **5**, 2298.

41. Muggleton, S., King, R., and Sternberg, M. J. E. (1992). *Protein Eng.*, **5**, 647.

42. Taylor, W. R. (1986). *J. Theor. Biol.*, **119**, 205.

43. Zvelebil, M. J., Barton, G. J., Taylor, W. R., and Sternberg, M. J. E. (1987). *J. Mol. Biol.*, **195**, 957.

44. Levin, J. M., Pascarella, S., Argos, P., and Garnier, J. (1993). *Protein Eng.*, **6**, 849.

45. Minsky, M. and Papert, S. (1988). *Perceptrons*. MIT Press, Cambridge, MA.

46. Sander, C. and Schneider, R. (1991). *Proteins: Struct. Funct. Genet.*, **9**, 56.

47. Bowie, J. U., Lüthy, R., and Eisenberg, D. (1991). *Science*, **253**, 164.

48. Gonnet, G. H., Cohen, M. A., and Benner, S. A. (1992). *Science*, **256**, 1443.

49. Hutchinson, E. G. and Thornton, J. M. (1994). *Protein Sci.*, **3**, 2207.

50. Smith, T. F. and Waterman, M. S. (1981). *J. Mol. Biol.*, **147**, 195.

51. Mehta, P. K., Heringa, J., and Argos, P. (1995). *Protein Sci.*, **4**, 2517.

52. Pascarella, S. and Argos, P. (1992). *Protein Eng.*, **5**, 121.

53. Heringa, J. (1999). *Comp. Chem.*, **23**, 341.

54. Heringa, J., Frishman, D., and Argos, P. (1997). In *Proteins: a comprehensive treatise, Vol. 1, principles of protein structure* (ed. G. Allen), pp. 171–277. JAI Press, New York.

55. Lüthy, R., McLachlan, A. D., and Eisenberg, D. (1991). *Proteins: Struct. Func. Genet.*, **10**, 229.

56. Langosch, D. and Heringa, J. (1998). *Proteins: Struct. Func. Genet.*, **31**, 150.

57. Unwin, N. (1993). *J. Mol. Biol.*, **229**, 1101.

58. Singer, S. J. (1962). *Adv. Protein Chem.*, **17**, 1.

59. Kyte, J. and Doolittle, R. F. (1982). *J. Mol. Biol.*, **157**, 105.

60. Argos, P., Rao, M. J. K.,and Hargrave, P. A. (1982). *Eur. J. Biochem.*, **128**, 565.

61. Sweet, R. M. and Eisenberg, D. (1983). *J. Mol. Biol.*, **171**, 479.

62. Cornette, J. L., Cease, K. B., Margalit, H., Spouge, J. L., Berzofsky, J. A., and DeLisi, C. (1987). *J. Mol. Biol.*, **195**, 659.

63. Von Heijne, G. (1992). *J. Mol. Biol.*, **225**, 487.

64. Klein, P., Kanehisa, M. I., and DeLisi, C. (1985). *Biochim. Biophys. Acta*, **815**, 468.

65. Rao, M. J. K. and Argos, P. (1986). *Biochim. Biophys. Acta*, **869**, 197.

66. Wang, J. and Pullman, A. (1991). *Biochim. Biophys. Acta*, **1070**, 493.

67. Eisenberg, D., Weiss, R. M., and Terwilliger, T. C. (1982). *Nature*, **299**, 371.

68. Eisenberg, D., Weiss, R. M., and Terwilliger, T. C. (1984). *Proc. Natl. Acad. Sci. USA*, **81**, 140.

69. Finer-Moore, J. and Stroud, R. M. (1984). *Proc. Natl. Acad. Sci. USA*, **81**, 155.

70. Vogel, H., Wright, J. K., and Jähnig, F. (1985). *EMBO J.*, **4**, 3625.

71. Esposti, M. D., Crimi, M., and Venturoli, G. (1990). *Eur. J. Biochem.*, **190**, 207.

72. Edelman, J. (1993). *J. Mol. Biol.*, **232**, 165.

73. Persson, B. and Argos, P. (1994). *J. Mol. Biol.*, **237**, 181.

74. Fariselli, P., Compiani, M., and Casadio, R. (1993). *Eur. Biophys. J.*, **22**, 41.

75. Rost, B., Casadio, R., Fariselli, P., and Sander, C. (1995). *Protein Sci.*, **4**, 521.

76. von Heijne, G. (1986). *EMBO J.*, **5**, 3021.

77. Boyd, D. and Beckwith, J. (1989). *Proc. Natl. Acad. Sci. USA*, **86**, 9446.

78. Sipos, L. and von Heijne, G. (1993). *Eur. J. Biochem.*, **213**, 1333.

79. Hartmann, E., Rapoport, T. A., and Lodish, H. F. (1989). *Proc. Natl. Acad. Sci. USA*, **86**, 5786.

80. Schirmer, T. and Rosenbusch, J. P. (1991). *Curr. Opin. Struct. Biol.*, **1**, 539.

81. Lupas, A., van Dyke, M., and Stock, J. (1991). *Science*, **252**, 1162.

82. Lupas, A. (1996). In *Methods in enzymology* (ed. R. F. Doolitle), Vol. 266, pp. 513–25. Academic Press.

83. Jones, D. T., Taylor, W. R., and Thornton, J. M. (1992). *Nature*, **358**, 86.

84. Jones, D. T., Orengo, C. A., and Thornton, J. M. (1996). In *Protein structure prediction: a practical approach* (ed. M. J. E. Sternberg), pp. 173–206. Oxford University Press, Inc., New York.

Chapter 7

Methods for discovering conserved patterns in protein sequences and structures

Inge Jonassen

Department of Informatics, University of Bergen HIB, 5020 Bergen, Norway

1 Introduction

The amount of available biomolecular data is exploding: the number of known protein sequences is increasing rapidly while the number of known structures is also increasing, though not as rapidly. A very useful observation in this situation is that common features among proteins can be used to group them into families, and then study the proteins on a family level. By a family we mean a set of proteins sharing some definite biological properties in terms of common function and/or structure, often implying that the proteins have evolved from a common ancestor, i.e. that they are homologous.

When studying a family, one can compare the sequences and structures (if known) of the proteins in the family in order to find what sequence or structure properties are shared by the family members and how these could explain the biological properties shared by the proteins in the family. A description of sequence properties is called a *sequence pattern*, and a description of structure properties is called a *structure pattern*. If a pattern is common to a family of proteins, it is called a *motif* for the family.

An example protein family is the set of proteins containing the classical zinc-finger DNA-binding domain. Most of the sequences in this family match the pattern C-x(2,4)-C-x(3)-[LIVMFYWC]-x(8)-H-x(3,5)-H, thus this is a sequence motif for this protein family. A sequence matches this pattern if it contains a C followed by 2–4 arbitrary letters followed by C and 3 arbitrary letters and one of L, I, V, M, F, Y, W, or C, and so on (this pattern notation is described in detail below). This particular pattern describes two cysteines and two histidines that are needed for coordinating the zinc ion in the classical zinc-finger domain, which means that this particular pattern has a direct biological interpretation. Additionally, the pattern can be used for classification, since not only do most sequences in the family match the pattern, but also very few sequences outside the family match the pattern. So, if one finds a match to this pattern in a new sequence, the chances are good that this new protein contains a zinc finger domain and binds to DNA.

A large number of patterns have been compiled and collected in different protein family databases. An example is the PROSITE database (1) which contains more than 1000 protein families and for most of these it gives a pattern which occurs in most of the sequences in the family. The patterns are regular expressions (like the zinc finger pattern given above) or profiles (position specific scoring matrices). Other databases also use local alignments, profiles, and Hidden Markov models (HMMs). We describe briefly some of these databases and how they can be used later in the chapter.

When one knows the structures of some of the proteins in the protein family under study, the structures can be compared and similarities described as patterns or motifs. In the same way that protein structures can be described at different levels (e.g. atom, residue, secondary structure, element level), structure patterns can describe structure properties at different levels. For example, one pattern could describe the packing of four alpha-helices and another pattern could describe the relative position of the cysteines and histidines in the classical zinc finger.

In this chapter we will use a very broad definition of patterns including both sequence and structure patterns and all the ways in which these can be defined. When going into more detail we will focus on sequence patterns which are of the regular expression type and on one particular type of structure patterns which describes packing of individual residues. In the following we discuss how existing databases of patterns can be used for analysing a new protein query sequence and how to assess the output of such a search. Later we describe different approaches to finding respectively sequence and structure patterns for a family.

2 Pattern descriptions

A very simple type of patterns is substring patterns—a sequence matches a substring pattern if it contains the substring (contiguous word in the sequence). For example, the substring pattern CDEC is matched by all sequences containing CDEC as a substring. This very simple type of patterns can sometimes be useful in the analysis of protein or nucleotide sequences. We will first define the concept of approximate pattern matching and then describe different generalizations of substring patterns.

2.1 Exact or approximate matching

When matching a sequence against a substring pattern one may allow for approximate matching. In this case, a sequence matches the pattern if it contains a substring approximately equal to the pattern. In practice, one defines a measure of distance between a pattern and a substring and sets an upper limit on the distance to be allowed. One simple way to measure the distance between two strings (or a pattern and a string) is to count the number of character changes needed to transform one into the other. This is called the number of mismatches or Hamming distance and can measure the distance between two strings only if they have equal length. For example, the sequence AGCDFCALKW approximately matches the substring pattern CDEC since the substring CDFC can be transformed into CDED by substituting the F with an E.

More general distance measures allow for insertion and deletion of characters in addition to substitutions. The edit distance between two strings (sequences) is the minimum number of single character insertions/deletions and substitutions needed to transform one into the other. For example, the sequence AGCDDALKW approximately matches the pattern CDED when one allows for an edit distance of more than one, but it does not match if one allows only for one mismatch. When comparing protein sequences, one may also penalize different substitutions differently since some amino acid replacements are found more often in equivalent positions in homologous (evolutionarily related) proteins. For example substitutions can be penalized using a substitution matrix, e.g. PAM-matrices (2) or BLOSUM matrices (3).

In order to find the edit distance between two sequences, one can use the dynamic programming algorithm (4) which can also be used when substitution matrices are used. For substring patterns, the matching problem is very similar to local pairwise alignment and database searching where speed-ups can be used, e.g. BLAST (5), Fasta (6).

2.2 PROSITE patterns

Using substring patterns and approximate matching one cannot specify that some pattern positions are compulsory (for instance, that there must be cysteines or histidines for binding zinc) while other positions are allowed to vary more freely. *Figure 1* illustrates this by showing four segments (substrings) containing the zinc finger motif and a consensus sequence (substring pattern) for the four. If this substring pattern is to match the four segments shown, one needs to allow up to 17 mismatches. A better way would be to allow no mismatches in the conserved positions, including the cysteines and histidines, and allow the remaining positions to be filled by any amino acid.

Some pattern languages allow a description of which amino acids are allowed at each position. For instance, in the PROSITE database (1) one uses a pattern language which is a subtype of regular expressions. Each pattern consists of a sequence of pattern elements that are of the following types:

(a) Single residue: matches one letter identical to the residue in the sequence, e.g. R matches an R in the sequence.

(b) Set of residues given in square brackets: matches any one sequence letter contained in the set, e.g. [KER] matches any one of K, E, or R in the sequence.

145

Local sequence alignment	Distance to consensus
EKPFACDFCGRKFARSDERKRHTKIHLRQKE	17
HKPFQCAICMRNFSRSDHLTTHIRTHTGEKP	1
HKPFQCRICMRNFSRSDHLTTHIRTHTGEKP	0
HKPFQCRICMRNFSRSDHLTTHIRTHTGEKP	0
HKPFQCRICMRNFSRSDHLTTHIRTHTGEKP	

Consensus sequence

Figure 1 Example of a local alignment of sequence segments containing the classical zinc-finger motif. The consensus sequence below shows for each position which amino acid is the most frequent, and to the right it is shown for each segment the number of mismatches between the segment and the consensus. Grey shading marks the positions conserved in all four segments. The positions of the conserved cysteines and histidines of the zinc-finger motif are underlined in the consensus sequence.

(c) Set of residues given in curly brackets: matches any one sequence letter not in the set, e.g. {KER} matches any letter except K, E, and R in the sequence.

(d) Wildcard x: matches any one letter in a sequence

Additionally:

(a) Single pattern elements can be followed by parentheses (i, j) which means that the sequence can contain between i and j (inclusive) letters each matching the preceding pattern element. For example, x(3,5) matches between 3 and 5 arbitrary sequence letters.

(b) The pattern can start with '<' meaning that the pattern should match from the beginning of a sequence.

(c) The pattern can end with '>' meaning that the pattern should match until the end of a sequence.

(d) The pattern elements are separated by hyphens '-'.

Consecutive pattern elements should be matched by consecutive sequence symbols, so for example the pattern C-x(2,3)-[DE] matches any sequence containing a C followed by two or three arbitrary letters followed by a D or an E.

2.3 Alignments, profiles, and hidden Markov models

Alternatives to regular expression type patterns are alignments, profiles, and HMMs. These can also be seen as generalizations of substring patterns. Effectively, for each position in the pattern, one now assigns a score (or a probability) to each of the 20 amino acids. Additionally, one assigns penalties (or probabilities) to insertions or deletions in each pattern position. Since alignments, profiles, and HMMs assign a score (probability) to a match to a sequence, they can be called *probabilistic*. Regular expression type patterns can be called *deterministic patterns*—they are deterministic in the sense that a sequence either matches or does not match the pattern (7).

146

A probabilistic pattern is normally constructed from a local alignment of a set of sequences from the family. A local alignment contains one (or more) segments from each sequence put on top of each other so as to align (put on top of each other) corresponding sequence positions. One may allow for insertion of special 'gap-characters' if needed to align corresponding positions. An example of a local sequence alignment (without gaps) is shown in *Figure 1* (see also Chapter 8). Normally, one would make an alignment of the parts of the sequences that are the most similar; for example in *Figure 1* the zinc-finger domains are aligned. Local alignments without gaps are used for example in the BLOCKS (8) and the PRINTS (9) databases. A sequence can be matched with (aligned to) an existing alignment using dynamic programming to optimize a score. A number of methods exist for scoring the match between an alignment of a single sequence letter to an alignment column and for penalizing gaps that may be inserted. Taking one alignment and a scoring scheme, one can make a scoring matrix giving a position specific scoring for each amino acid and also position specific gap penalties. Such descriptions are called profiles and were initially suggested by Gribskov *et al.* (10). An alternative approach is to use the framework of Hidden Markov Models (HMMs) (11), which, in the way used in sequence analysis, provides a scoring scheme similar to that of profiles, but based on probabilities (see *Chapter 4*).

This means that the information contained in an alignment is often represented as a weight matrix, a profile, or a HMM which specifies for each column in the alignment a score for each of the 20 amino acids when it is aligned with this position. The scores can be calculated from the distribution of amino acids in the column as well as using external information from substitution matrices or Dirichlet mixtures (12). Schemes have been developed to weight the sequences so as to adjust for biases in the input set (13).

PROSITE is also using profiles as a supplement to regular expression type patterns, since for some families it is not possible to define one single regular expression type pattern which matches all family sequences while avoiding matches in unrelated sequences. In such cases it can be possible to define a profile matching the sequences in the family with a higher score than all (known) sequences outside the family. This may be possible since profiles have more expressive power than the deterministic patterns as they can assign different scores to each amino acid when matched to each pattern position and also position-specific gap-penalties. On the other hand, profiles and other probabilistic patterns contain many more parameters to be estimated, and to estimate their values one needs a large number of examples (family members). Also, if one is to learn patterns from noisy examples (including unrelated sequences), the large number of parameters makes it easier to adapt the patterns to match unrelated sequences. In the context of learning patterns from noisy examples, therefore, the deterministic patterns can be more appropriate. Also, deterministic patterns are very simple, mathematically pure, and the human mind finds them easy to interpret. Whichever patterns one choose to use, an apparent problem is how to assess the patterns, to decide between alternative patterns which is the best one, and to find whether the identified pattern could be the result of chance.

2.4 Pattern significance

When the quality of a pattern is to be assessed, it depends on what is the purpose of the pattern. One typical application of patterns is classification; that is, the patterns are to be used for discriminating between family members and non-members. Another is to find patterns that describe biologically important features. Below we discuss some ways of assessing the quality of patterns with respect to each of the two applications (see also ref. 7).

2.4.1 Characterization

Patterns can be used to describe biologically important features of the proteins, that is, one wants to describe which features are compulsory in order for the protein to belong to a particular family and which features are optional. If one has available a set of sequences (or structures) from the same family, these can be compared, and it can be determined which residues are conserved through evolution and therefore likely to be important to the proteins' function and structure. If the available proteins have undergone little evolution since their last common ancestor (for example if their sequences are 90% identical), it will not be easy to find which of the conserved residues are most important for the biological function of the proteins. Therefore one should try to collect proteins that are as diverse as possible while avoiding inclusion of unrelated proteins.

Having developed a pattern conserved in a set of sequences, one should find whether such a pattern is likely to be conserved by chance and therefore not necessarily biologically important. One common method is to calculate an estimate of the probability that a set of random sequences (equal in number and lengths to the sequences under analysis) would share a pattern of the same strength as the identified pattern, as a result of chance. If this probability is very low, one has a better reason for believing that the identified pattern has some biological meaning. When evaluating the significance of the discovered pattern, one should also take into account the number of patterns that have been considered in the pattern discovery phase (see e.g. ref. 14). For example, a pattern with probability 10^{-6} is expected to be found if one million patterns are considered.

An alternative to calculating pattern probabilities is to measure the patterns' information content (15). The higher the information content the pattern possesses, the less likely a random sequence is to match it. The measure was designed for ranking patterns matching the same number of sequences. Using the principle of minimum description length (MDL) from machine learning (16), this has been extended to also take into account the number of sequences matching each sequence (17).

An alternative approach is to do a series of pattern discovery experiments on sets of sequences with characteristics similar to the sequences in which the patterns were found. The sequences should be chosen so that they share no significant patterns. The result of the pattern discovery experiments will give information about what type of patterns can be found by chance. For example, one can repeat x times: shuffle all the sequences, and check which patterns can be found to match at least the same number of (the shuffled) sequences as the pattern under analysis. An advantage of this approach is that in assessing the 'background probability' one can use sequences which have the same characteristics as the original similar sequences (local sequence composition, etc.) for example by using special shuffling operations.

When evaluating discovered patterns, it is important to take into consideration whether (some of) the sequences under analysis are very closely related. When calculating the probabilities, the model normally assumes that the sequences are independently generated by some probabilistic model and the shuffling would normally also extinguish any close similarities between the sequences. If some of the sequences are very similar, they will contain many common patterns, and any pattern matching one of them will probably match all, and is therefore likely to be deemed as more significant. A scoring scheme taking this into account, has been proposed (18).

2.4.2 Classification

When a pattern is to be used for classification, it should ideally match all family members and no other sequences. Most often, however, the pattern fails to match some member sequences (called false negatives), and it may match some sequences outside the family (false positives). For an illustration, see *Figure 2*. The fewer false negatives, the more sensitive the pattern is said to be, and the fewer false positives, the more specific it is. Ideally, a pattern should have zero false positives and negatives.

An estimate of the number of matches in a sequence database can be found by multiplying the probability that one random sequence matches the pattern by the number of sequences in the database. In order to calculate the probability we assume that random sequences are generated using a specific probabilistic model. Sternberg (19) did this for all the patterns in the PROSITE database and showed a clear correlation between the expected number of false positives and

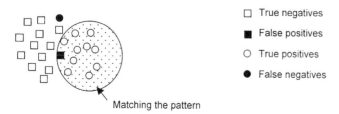

☐ True negatives
■ False positives
○ True positives
● False negatives

Matching the pattern

Figure 2 Illustration of the concepts of true positives, true negatives, false positives, and false negatives. The circles are the family members and the squares are non-family members. A pattern matches the encircled objects, and the status of each object is shown by its colour (unfilled means 'true' and filled means 'false').

the actual number, i.e. the number of unrelated sequences in the SWISS-PROT database (20) matching the pattern.

Denoting the number of true positives (sequences in the family matching the pattern) by TP and the number of false negatives by FN, the *sensitivity* of a pattern (21) can be defined as

$$Sensitvity = TP/(TP+FN)$$

and measures of how big a proportion of the family sequences are 'picked up by' (matched by) the pattern. Similarly, the *specificity* of the pattern can be defined as

$$Specificity = TN/(TN+FP)$$

(where TN and FP are respectively the number of true negatives and the number of false positives) which measures of how big a proportion of the sequences outside the family are not matched by the pattern. Yet another useful number is the positive predictive value (PPV) which says how big a proportion of the sequences matching the pattern are actually in the family,

$$PPV = TP/(TP+FP)$$

The value range for all three is from zero to one, one being the best possible. When evaluating patterns to be used for classification, one needs to use more than one of the measures. This can be illustrated by two degenerate cases, (1) the empty pattern matching any protein, and (2) a pattern matching one single protein being member in the family. Pattern (1) has perfect sensitivity, but very bad specificity and PPV, while pattern (2) has perfect specificity and PPV, but bad sensitivity. For a concrete example of the use of these equations, see below. In practice one often needs to make a trade-off between sensitivity and specificity when choosing which pattern to use for a family. One way to evaluate a probabilistic pattern's ability to discriminate between family members and other sequences is to find a cut-off on the score that gives the same number of false positives and false negatives. Tatusov *et al.* (22) evaluated alternative ways of finding weight matrices from local ungapped alignments using this approach.

2.4.3 Discussion

Often the patterns that describe biologically important features will also be good for classification purposes and *vice versa*. However, it is possible that a pattern that gives perfect discrimination can be derived and yet lacks any biological interpretation. Also, it may be that the features described by a pattern are important in the family, but not unique to the family, so that it is not specific enough to be used for classification purposes.

2.5 Pattern databases

A number of different databases for storing information about protein families and motifs have been established during the last ten years. They differ in a number of ways. Firstly, they differ in how they represent the patterns for each family. Secondly, some of them are constructed manually (both sequence group-

Table 1 Summary information about some protein family databases

Database	Pattern type	URL
PROSITE	Reg.exp and profiles	Http://expasy.hcuge.ch/sprot/prosite.html
BLOCKS	Blocks (ungapped local alignments)	Http://www.blocks.fhcrc.org/
Prints	Blocks	Http://www.biochem.ucl.ac.uk/bsm/dbbrowser/ PRINTS/
Identify	reg. exp	Http://motif.stanford.edu/identify/
Pfam	HMMs	Http://www.sanger.ac.uk/Pfam/

ing and pattern definition) while others are made (to a varying degree) automatically. For some summary information about a few family databases, see *Table 1*.

2.5.1 PROSITE

One of the most widely used databases is PROSITE (1), in which, or each family, one or several patterns and/or profiles are given in the format described above (*Section 2.2*). Profiles are described using the generalised profile syntax (23). Statistics are given which describe the patterns' ability to discriminate between family members and other sequences given in the SWISS-PROT protein sequence database (20), in the form of the number of false positives and false negatives. Also, a number of unknowns is given which is the number of sequences in SWISS-PROT which match the pattern, but for which it is not yet known whether it belongs to the family or not.

Figure 3 shows a PROSITE entry giving a signature pattern (motif) for the actinin-type actin-binding domain. We will explain the most important (for our purpose) parts of the entry. First, the ID and AC lines give, respectively, the name and the accession number of the PROSITE entry. The pattern is given on the PA line (the pattern can continue over several PA lines, then end of the pattern being marked by a period sign). The NR lines give statistics about the pattern's discriminatory power with respect to the SWISS-PROT database. The first NR line says that the statistics are with respect to release 35 of SWISS-PROT, which has 69 113 sequence entries. The next NR line gives the number of matches of different categories (true positives, false positives, false negatives, etc.) in the SWISS-PROT database. Each is given as $x(y)$ meaning that there are x matches to the pattern in y different sequences.

The DR lines give references to the corresponding SWISS-PROT entries both by their names and accession numbers. Each reference is on the form 'AC, ID, Status;' where Status is one of T (true positive), N (false negative), F (false positive), P (partial), and ? (unknown). Finally, the 3-D line gives names of PDB (Protein Data Bank) entries containing structures of proteins in the family, and the DO line gives the accession number of the entry in the PROSITE documentation part corresponding to this entry.

151

```
ID    ACTININ_1; PATTERN.
AC    PS00019;
DT    APR-1990 (CREATED); NOV-1997 (DATA UPDATE); NOV-1997 (INFO UPDATE).
DE    Actinin-type actin-binding domain signature 1.
PA    [EQ]-x(2)-[ATV]-[FY]-x(2)-W-x-N.
NR    /RELEASE=35,69113;
NR    /TOTAL=55(46); /POSITIVE=35(28); /UNKNOWN=0(0); /FALSE_POS=20(18);
NR    /FALSE_NEG=0; /PARTIAL=1;
CC    /TAXO-RANGE=??E??; /MAX-REPEAT=2;
DR    P12814, AAC1_HUMAN, T; P35609, AAC2_HUMAN, T; Q08043, AAC3_HUMAN, T;
DR    Q99001, AACB_CHICK, T; Q90734, AACN_CHICK, T; P20111, AACS_CHICK, T;
DR    P05094, AACT_CHICK, T; P05095, AACT_DICDI, T; P18091, AACT_DROME, T,
DR    P21333, ABP2_HUMAN, T; P11533, DMD_CHICK , T; P11532, DMD_HUMAN , T;
DR    P11531, DMD_MOUSE , T; P19179, FIMB_CHICK, T; P54680, FIMB_DICDI, T;
DR    P32599, FIMB_YEAST, T; P13466, GELA_DICDI, T; Q14651, PLSI_HUMAN, T;
DR    P13796, PLSL_HUMAN, T; Q61233, PLSL_MOUSE, T; P13797, PLST_HUMAN, T;
DR    Q63598, PLST_RAT  , T; Q00963, SPCB_DROME, T; P11277, SPCB_HUMAN, T;
DR    P15508, SPCB_MOUSE, T; Q01082, SPCO_HUMAN, T; Q62261, SPCO_MOUSE, T;
DR    P46939, UTRO_HUMAN, T;
DR    P11530, DMD_RAT   , P;
DR    P13688, BGP1_HUMAN, F; P06731, CCEM_HUMAN, F; P31997, CGM6_HUMAN, F;
DR    P40782, CYP1_CYNCA, F; P10474, GUNB_CALSA, F; P40199, NCA_HUMAN , F;
DR    P11462, PBG1_HUMAN, F; P11463, PBGC_HUMAN, F; P11464, PBGD_HUMAN, F;
DR    Q00887, PSGB_HUMAN, F; P35853, PUR1_LACCA, F; P14410, SUIS_HUMAN, F;
DR    P23739, SUIS_RAT  , F; P28668, SYEP_DROME, F; P07814, SYEP_HUMAN, F;
DR    P42954, TAGH_BACSU, F; P09301, UL07_VZVD , F; P52583, VGR2_COTJA, F;
3D    1KSR; 1DRO; 1BTN; 1MPH;
DO    PDOC00019;
```

Figure 3 Example of a PROSITE entry taken from release 14.0 (November 1997). See text, for a detailed explanation.

Referring to the pattern quality measures discussed earlier, this particular pattern has sensitivity $28/(28 + 0) = 1$, specificity $(69\,113 - 28\text{–}20)/(69\,113 - 28)$ $= 0.9997$ ($69\,113 - 28$ is the number of sequences in SWISS-PROT outside the family), and PPV $28/(28 + 18) = 0.608$. The last number means that if a sequence from SWISS-PROT matches the pattern, the probability that it belongs to the family is 60.8%. For this particular application, the PPV measure seems more meaningful than specificity since the number of false positives do not affect the measure of specificity very much as the number of true negatives most often will be very much bigger than the number of false positives. Each entry in the PROSITE documentation part gives a description of the family and explains the biological significance of the signature patterns. It also gives literature references and one or several experts that can be contacted for more information about the family. The PROSITE database is largely maintained manually; new families, profiles and patterns are carefully scrutinized.

2.5.2 BLOCKS

Another protein family database is BLOCKS (8) which contains the same families as PROSITE, but instead of giving patterns or profiles, it gives a set of blocks for each family. A block is an ungapped local multiple alignment. The database is constructed fully automatically. For each family the member sequences (as given in PROSITE) are subjected to pattern discovery and local alignment methods. The blocks can also be linked in chains. It is recommended that a query sequence is matched against both PROSITE and BLOCKS, because even though they describe the same families, the patterns and the blocks in a sense have complementary strengths when used as classifiers.

2.5.3 PRINTS

The PRINTS database is a collection of fingerprints and is constructed semi-automatically (9). A fingerprint is defined as a list of motifs, each motif being a local ungapped alignment. The fingerprints are made by first manually making an alignment of some family members and then iteratively scanning a database of protein sequences, adding new members, updating the fingerprint, scanning again until convergence (no new family members are found). Each entry in PRINTS, gives the local alignments corresponding to the motif and information about partial matches etc. On the website of PRINTS, a tool FingerPRINTScan, is available for scanning a query sequence against the database. The output can be visualized showing the position of the motif matches in the query sequence.

2.5.4 PFam

Pfam is a database of multiple sequence alignments and HMM-profiles of protein domains (24). It is partly manually curated. Seed alignments for each family are made semi-automatically, and these are extended using HMM methods. Version 3.1 (August 1998) contains 1313 families.

On the Web site there are available tools for matching a query sequence against the database. Also, a special database SWISSPFam is made which shows the domain organization (according to Pfam) or the sequences in SWISS-PROT and TrEMBL.

2.5.5 Identify

Identify (25) is a database of patterns of the same form as used in PROSITE but without flexible length wildcards. It contains patterns for the families in the BLOCKS and PRINTS databases. The patterns were constructed from the ungapped alignments (blocks) in the BLOCKS and PRINTS databases by using a pattern finding program (EMOTIF). For each block, there can be several patterns so that each pattern matches a subset of the sequences. The patterns were generated to have a certain specificity (calculated as the probability that a random sequence matches the pattern by chance, cf. ref. 19), and patterns were generated for different specificity levels. The World Wide Web server allows the user to input a query sequence which then is matched against the pattern collection and the user is given the list of matching patterns together with links to the corresponding BLOCKS and PRINTS database entries.

2.6 Using existing pattern collections

Most of the family databases are available on the World Wide Web, and they also provide on-line tools for (1) matching a sequence against the pattern in the database, and sometimes (2) matching a new (user-defined) pattern against a sequence database. For example, on the PROSITE website, the search engine

ScanProsite that can be used for both (1) and (2) with regular expression type patterns. For scanning against the profiles in PROSITE, the tool ProfileScan can be used. The ScanProsite program does not return any information about the probability that the match between the sequence and the pattern could be by chance, and it does only allow for exact matching between the pattern and the sequence.

Another very useful tool is PdbMotif (26), which takes as input a protein structure and finds all matches between the protein's sequence and patterns in the PROSITE database. It generates a script that can be input to RasMol (27) to highlight the pattern matches. PdbMotif also outputs a probability that the sequence should match the PROSITE pattern by chance, which can help to identify possible false positives.

3 Finding new patterns

Analysing a set of proteins believed to be related one may want to find a motif describing the set. The goal may be to find which features are common to the proteins under study helping to better understand the relationships between sequence, structure, and function of the proteins. Motifs can also help to identify new possible family members. We may also want to find motifs to be included in protein family databases where the motif may later be used for classification.

One may develop motifs semi-manually using knowledge about the proteins under study and for example alignments generated either manually or automatically using a multiple (local or global) sequence alignment program. If the alignment programs were guaranteed to find the correct alignments, this approach would be all you needed. However, multiple alignments are often difficult to obtain and interpret. Therefore, in many cases, direct methods for finding motifs directly from the sequences or structures can lead to better results. In such cases, the motifs found can also be used to guide the alignment of the sequences or structures in the family.

3.1 A general approach

There exist a large number of methods for the discovery of patterns from protein (and DNA) sequences. A survey of these is given by Brazma *et al.* (7) who propose a general three-step approach to pattern discovery:

(a) Choose a **solution space**, i.e. set of patterns that the method potentially can discover.

(b) Define a **fitness function** reflecting how well a pattern fits the input sequences.

(c) Develop an **algorithm**, which given a set of input sequences, returns the pattern(s) from the solution space with high (highest) fitness.

The different methods can be classified according to their approach to each of the three steps. This three-step approach can be used for the discovery of all types of sequence patterns (including regular expression type patterns, profiles, and Hidden Markov models) and also for the discovery of motifs in protein structures. An additional criterion is whether a method is guaranteed, for any input set, to find the best (as measured by the fitness function) pattern in the solution space. In the following sections we will illustrate the three steps approach by describing in some detail some representative methods, especially focusing on the Pratt and SPratt methods.

3.2 Discovery algorithms

Previous sections have discussed in some detail different solution spaces and principles of fitness functions. Here we discuss in some more detail the third step, namely that of finding the patterns in the solution space that have high, and possibly the highest, fitness. Brazma et. al. (7) identified two main algorithmic approaches to this problem:

(a) Pattern Driven (PD) methods.

(b) Sequence Driven (SD) methods.

3.2.1 Pattern driven methods

In the simplest form, PD methods enumerate all patterns in the solution space calculating each pattern's fitness so that the best ones can be output. For example, one step of the algorithm by Smith *et al.* (28) works by enumerating all patterns of the form A1-x(d1)-A2-x(d2)-A3 having single amino acid symbols and d values up to some maximum (e.g. 10). For each pattern, the number of matching input sequences were counted, and the patterns matching the most sequences were subjected to further analysis.

Some methods have mechanisms for avoiding looking at all patterns in the solution space, for example by not searching parts of the solution space that cannot possibly contain patterns scoring higher than already analysed patterns. Often heuristics are used to discard parts of the solution space that are unlikely to contain high scoring patterns. Examples of this will be seen when the Pratt method is described in some more detail below (see *Figure 4*).

3.2.2 Sequence driven methods

These methods work by comparing (normally pairs of) the input sequences and expressing local similarities as patterns. By repeating this for different pairs of sequences, or by subjecting the patterns themselves to pairwise comparison, patterns are gradually built up that match many or all of the input sequences. Some of these methods are very closely related to methods for sequence alignment.

For example, the pattern discovery method proposed by Smith and Smith (29) first computes the similarity between all pairs of input sequences. The most similar pair is input to a local pairwise alignment method that is based on dynamic programming (30). Instead of producing a pairwise alignment, this algorithm outputs a pattern common to the two sequences. The pattern is basic-

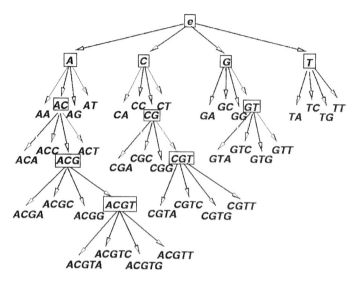

Figure 4 Example of a pattern-driven discovery method applied to a set of four sequences: ACGTT, TACGTA, GCACGT, and TTACGTAA. The solution space consists of all words over the DNA alphabet (no wildcards or group characters), and patterns are sought that match all four sequences. Patterns matching all four are boxed, and only these are extended. The longest pattern found to match all four sequences is ACGT, The 'e' at the top is the empty pattern which matches any sequence.

ally of the PROSITE type, but the set of possible amino acid groups is limited and given by an Amino Acid Class Covering (AACC) hierarchy. The two sequences aligned are now replaced by their common pattern, and the procedure is repeated until there remains only one pattern matching all of the input sequences. For an example, see *Figure 4*. Note that this approach is very analogous to the progressive multiple alignment methods used for example by Thompson *et al.* (31) and Taylor (32).

4 The Pratt programs

The Pratt programs take as input a set of (un-aligned) sequences and finds patterns matching at least a (user-defined) minimum number of the sequences. The patterns are of the type used in PROSITE and can include both character groups and flexible length wildcards. For instance, Pratt is able to automatically rediscover the pattern C-x(2,4)-C-x(3)-[LIVMFYWC]-x(8)-H-x(3,5)-H from the set of over 300 classical zinc-finger-containing protein sequences in SWISS-PROT in less than a minute on a modern workstation. No prior information about the pattern is given to Pratt and the sequences are input unaligned.

The user can input constraints on the patterns to be considered, effectively defining the solution space to be used (cf. three-step approach). The user also chooses how many of the input sequences a pattern should match, in order to be output. We will call any pattern matching at least this number of sequences, a conserved pattern. Pratt uses a two-step search for finding conserved patterns from the chosen solution space having maximum fitness. The fitness is normally defined as the information content of the pattern (see below).

In the following sections we give a practical guide to how Pratt should be used and then some details about the algorithms used in Pratt. For more detailed technical descriptions, see the original papers (15, 33) and the Pratt home page on the World Wide Web: http://www.ii.uib.no/~inge/Pratt.html.

4.1 Using Pratt

When using Pratt from a command line environment, it is started using the command

```
pratt <format> file [options]
```

where `format` is one of `fasta` or `swissprot` and `file` is a file containing the sequences in the indicated format. Optionally, one can specify options on the command-line. If no options are given, a menu appears. *Figure 5* shows the menu of Pratt when run on a file MUTT containing 26 unaligned sequences. The menu can be used to choose values for a number of parameters and also to obtain help. The parameters fall within a few main classes:

(a) **Pattern conservation**. Using options CM (respectively, C%) one can set the minimum number (respectively, percentage of the input sequences) of sequences a pattern should match.

(b) **Pattern restrictions**. A number of parameters are used to constrain the patterns to be considered. For example PL can be used to set the maximum length of a pattern, PN to set the maximum number of non-wildcard symbols, PX to set the maximum length of a wildcard region. For constraining the flexibility to be allowed in wildcard regions, FN is used to set the maximum number of flexible wildcards and FL to set the maximum flexibility of a

```
Pratt version 2.2:  Analysing 26 sequences from file MUTT

  PATTERN CONSERVATION:              SEARCH PARAMETERS:
  CM: min Nr of Seqs          26     G:  Pattern graph from        seq
  C%: min Percentage       100.0
                                     E:   Search greediness         3
  PATTERN RESTRICTIONS :             R:   Pattern Refinement        on
  PP: pos in seq             off     RG: Generalise                off
  PL: max Length              50
  PN: max Nr of Symbols       50     OUTPUT:
  PX: max Nr of x's            5     OF: Output filename   MUTT.26.pat
  FN: max Nr of flex.          2     OP: PROSITE pat notation        on
  FL: max Flexibility          2     ON: max nr patterns             50
  BI: Symbol File            off     OA: max nr alignments           50
                                     M:   Match summary              on
  BN: Initial Search          20     MR: Ratio                       10
                                     MV: Vertical summary           off
  PATTERN SCORING:
  S:  Scoring               info

  X: eXecute program      Q: Quit    H: Help

  Command:
```

Figure 5 Pratt menu. The options are explained in the text.

wildcard. Note that the PX constraint also applies to patterns to be found during the initial search. For some examples, see *Table 2*. Also, one can choose which pattern symbols should be used during initial pattern search and during pattern refinement using options BI (BF) and BN.

(c) **Pattern scoring**. By default patterns are scored by their information content. Optionally one can use a scoring function derived from the Minimum Description Length (MDL) principle. Patterns can also be ranked by their positive predictive value (PPV)—in this case the name of a file containing a sequence database in flat file format must be given using option SF.

(d) **Search parameters**. By default, the shortest sequences will be used for deriving a pattern graph (see below). Optionally this can be generated instead from a special query sequence or from an alignment, in which case only patterns that match the query sequence respectively are consistent with the alignment, will be considered in the search.

(e) **Output format**. Filename for output can be chosen using option OF, format of patterns using option OP, if pattern matches should be shown (option OA), etc.

(f) **Help and control**. Help can be obtained by typing in option H, the search started by using option X, or abandoned using option Q.

All parameters that can be set using the menu can alternatively be set from the command line by adding '-<menu option> <value>' to the command. For instance

```
pratt fasta seqs -cm 20
```

tells Pratt to analyse the sequences in the file `seqs` (fasta format) to find patterns matching at least 20 of the sequences. When command line options are used, the menu will not appear unless the option `-menu` is used.

For example, if the file c2h2 contains the sequences (in fasta format) of the proteins in the classical c2h2 zinc finger family, these can be analysed using the command:

```
pratt fasta c2h2 -px 15
```

Table 2 The table shows some example patterns, and for each example, the minimum values to be used for some Pratt parameters if the pattern is to be discovered. For example, in order to discover the bottom-most pattern, one needs to increase the value of the PX parameter to at least 12

Pattern	Minimum parameter values				
	PL	PN	PX	FL	FN
C-x(3)-C	5	2	3	0	0
C-x(2,5)-C	7	2	5	3	1
C-C-D-E-x(7)-C	12	5	7	0	0
C-x(2,4)-C-x(12)-H-x(3,5)-H	25	4	12	2	2
Default values	50	50	5	2	2

using the option -px 15 to allow long wildcards (if we want to re-discover the known motif, we need to set PX to at least 12 since during the initial pattern search, a spacing of 12 is needed between the last conserved cysteine and the first histidine). Also, if one wants to find all patterns matching a minimum 90% of these sequences, one can use the command

```
pratt fasta c2h2 -px 15 -c% 90
```

4.2 Pratt: Internal search methods

The search for conserved patterns is done in two phases:

(a) **Initial pattern search**. Search for patterns having only single character elements and wildcards (possibly of flexible length). For example, in this phase the pattern Pratt can discover that the pattern A-x(4)-D-x-E is conserved. Optionally, group characters can be allowed also in this step, at the cost of increased computing time.

(b) **Pattern refinement**. Take each of the best patterns from phase 1, collect the matching sequence segments and check if wildcard positions can be replaced by group characters so that the pattern remains conserved. Also, the pattern may be extended to the right (but not to the left) in this phase. For example, the pattern above can now be refined to

```
A-x-[KER]-x(2)-D-[ILV]-E-x(4)-[KR].
```

4.2.1 Initial pattern search

In the first version of Pratt (15), in the first phase a search tree containing all patterns to be considered, is explored. The label of the root node is the empty pattern. A node having the pattern P as label has children with labels $P\text{-}x(i,j)\text{-}A$, that is P extended with a wildcard region (empty if $i-j=0$) and a single character A, for all allowable values of $i, j,$ and A.

The search starts at the root of the tree. At each step of the search a node in the tree having label P is considered. It is assumed that P is conserved. Then, all children of P are generated, and for each of these it is checked whether the corresponding pattern is conserved (whether it matches the minimum number of sequences). If a pattern is conserved, this is recursively analysed using the same procedure. If no extension of P is conserved, then, depending on the score of P, it is included in the list of the best patterns that are subsequently input to the refinement algorithm.

In the second version of Pratt (33), instead of considering the full tree of all patterns in the solution space, a pattern graph is used to define the set of patterns to be considered. The pattern graph is a node- and edge-labelled directed acyclic graph. The nodes are labelled with non-wildcard pattern elements and the edges have as labels wildcard lengths. A path in the graph defines a pattern and from this pattern, more generalized versions will be generated. An example of a pattern graph is shown in *Figure 6*.

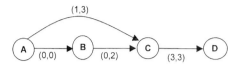

Figure 6 Example of a pattern graph. The paths in the graph define the patterns A-B-x(0,2)-x(3,3)-D, A-B-x(0,2)-C, A-B, B-x(0,2)-C-x(3,3)-D, B-x(0,2)-C, C-x(3,3)-D, A-x(1,3)-C-x(3,3)-D, A-x(1,3)-C.

The initial search explores all patterns that can be derived from paths in the pattern graph and that are contained in the class of patterns defined by the user. The search is focused on finding only the highest scoring patterns. Branch-and-bound techniques are used to avoid considering parts of the search space that cannot possibly contain patterns with higher scores than already identified patterns.

Also, heuristics have been implemented that effectively avoids exploring search paths unlikely to produce patterns scoring higher than patterns already found. The user can adjust the greediness of the search. Setting the E parameter to zero gives non-heuristic search (guaranteed to find highest-scoring patterns), setting E to 1 gives the same guarantee in cases where no flexibility is allowed in wildcard regions, and E values above 1 gives increasingly greedy search. The default value is 3. The more greedy the search, the faster it will be, and the more likely Pratt is to not find the highest scoring patterns. Experiments have shown that $E = 3$ gives a good compromise between speed and accuracy for protein sequences, while for DNA a lower value should be used (for instance, $E = 1.5$).

4.2.2 Pattern refinement

During refinement, each position in the fixed-length wildcard regions (excluding regions x(i, j) where j > i) is analysed and it is checked whether replacing the wildcard with an allowed group character (the allowed groups are given as a list) so that the resulting pattern remains conserved. It might be that there are several wildcard positions that can be replaced by group characters, so that replacing any one of these gives a conserved pattern, but if all are replaced simultaneously, the resulting pattern will not be conserved. As there are exponentially many (in the number of positions that can be replaced) subsets of replacements that could be done, it is computationally expensive to consider all. Therefore, in the second version of Pratt (33), a heuristic refinement algorithm is used. The degree of greediness is adjusted using the same E-parameter as for the initial search.

4.2.3 Block data structure

To find all matches to each pattern quickly, a special data structure initially proposed by Neuwald and Green (14) is used. Assume that the input sequences are $S = \{ S1, S2, \ldots, Sn \}$. For some fixed parameter w (typically 50), let B be the set of all w-segments (consecutive substrings of length w) from all of the sequences in S. Also construct w-segments in the end of the sequences by appending to

Input sequences:

```
MDNVVDPWYI
MANVEKPND
MMHIKSLPHAHH
```

The corresponding 5-segments:

```
B={MDNVV,DNVVD,NVDDP,VVDPW,VDPWY,DPWYI,PWYI-,WYI--,YI---,I----,
   MANVE,ANVEK,NVEKP,VEKPN,EKPND,KPND-,PND--,ND---,D----,
   MMHIK,MHIKS,HIKSL,IKSLP,KSLPH,SLPHA,LPHAH,PHAHH,HAHH-AHH--,HH---,H----}
```

Figure 7 Example of the set B of w-segments made for a set of sequences. The B segments are used in the block data structure.

each sequence *w-1* dummy symbols '-'. For an example of the *w*-segments made for a set of sequences, see *Figure 7*.

Now, for each amino acid symbol *a*, and for each *i* between *1* and *w*, construct the set $b_{i,a}$ that is the set of all *w*-segments having character *a* in position *i*. These sets can be used to quickly find the set of *w*-segments matching any pattern considered by Pratt not having length exceeding *w*. For instance, the set of segments matching A-x(2)-B is $b_{1,A} \cap b_{4,B}$. In the recursive search, the set of segments matching P is used together with the block data structure to find the segments matching each extension P-x(i,j)-A of P. For a more detailed description, see Jonassen *et al.* (15).

4.3 Scoring patterns

Pratt can score discovered patterns using different fitness functions. By default the patterns are evaluated by their information content (15). This measure depends only on the pattern itself and is only appropriate for ranking patterns matching the same number of sequences. The information content of a pattern is a measure of the information gained about an unknown sequence when one is told that the sequence matches the pattern. It increases with the number of single character elements (any one single character contributing the same) and with the number of group character elements (more ambiguous contributing less). A penalty is subtracted for flexible wildcard regions, more specifically *x(i,j)* is penalized by *c(j-i)* where *c* is a parameter whose default value is *0.5*.

As an alternative, Pratt can score the discovered patterns using the minimum description length (MDL) principle based fitness measure described by Brazma *et al.* (17). This assigns a score to a pattern that depends on the pattern's information content and on how many sequences it matches. The fitness measure definition contains some parameters that can be used to slant the optimum towards strong patterns matching few sequences or towards weaker patterns matching many sequences. Brazma *et al.* (17) used the measure in a method for simultaneously finding subfamilies and patterns in a set of unaligned (and un-labelled) sequences. It was required that each pattern matches all sequences in one of the subfamilies. The method uses Pratt to find patterns matching different sized subsets of sequences. Next it selects in a greedy fashion a collection of the patterns that cover the input sequence and has a high fitness value.

If the aim is to find patterns to be used for classification, Pratt can evaluate the discovered patterns by their positive predictive value (PPV, see *Section 2.4.2, Classification*). It is assumed that the sequences under analysis are all in the SWISS-PROT database (20). The number of false positives for each pattern is found by matching the patterns against the SWISS-PROT database that must be locally available in flat file format.

5 Structure motifs

Finding recurring patterns (motifs) in protein structures help to better understand the rules underlying the formation of protein structures. Since structure is better conserved during evolution than sequence, structural similarities can also help to identify remote evolutionary relationships. Structure motifs can help in approaching the structure prediction problem and in assigning function to proteins. Structure motifs can represent common structural features at different levels. For example, they can represent packing of secondary structure elements, local packing of residues, and atom coordinates of binding atoms in active site (or ligand binding) residues. Structure motifs describing functional sites in proteins, have been developed by, for example, Wallace *et al.* (34). They call their motifs 'templates' and suggest that they can be used for finding functional sites in proteins. They have also developed a database PROCAT of such templates (35) which allows the user to search for 3-D enzyme active sites in a protein structure.

In order to find recurring patterns in protein structures one can use methods for the comparison of protein structures. A number of such methods have been developed, most of them for comparing pairs of structures, but also some for multiple structure comparison. The methods differ in what similarities they are able to find. Some represent the structures as composed of secondary structure elements (alpha helices, beta strands, and loops) and have provided methods for finding patterns of conserved patterns at this level. Other methods find patterns of residues (or atoms) that have similar configurations in space. Brown *et al.* (36) gives a survey of a large number of different methods focusing on their way of representing similarities. An important difference between methods is whether they require matched elements to be in the same order along the proteins' primary structure. A number of different methods have been used, including extensions of the dynamic programming algorithms used for pairwise sequence (37), use of graph-theoretic methods (38), and methods from computer vision (39).

Also, some more direct methods for structure motif discovery have been suggested. Here we will describe the SPratt program, a more detailed description of which can be found in (40).

5.1 The SPratt program

The idea behind the SPratt program was to use the Pratt method developed for sequence motif discovery, to discover structure motifs. This can be done by encoding structural features in the form of strings and input these to Pratt producing patterns common to the strings. Next one needs to check if the patterns found in the structure description strings correspond to similarities between the structures. The encoding of structural properties that we adopted was one described by Karlin and Zhu (41). In their method, they make one string per residue in each structure. The strings contain information about the spatial neighbourhood of each residue. Karlin and Zhu describe alternative methods for making these strings, and we chose one of them to be used in SPratt.

Overview of the method:

1. For each residue a, we make two neighbour strings Ca and Na. The string Ca starts with a followed by all the residues C-terminal to a, whose spatial distance to a is below a user-chosen threshold $dmax$. The residues are ordered in N-to-C chain order. Analogously, the string Na starts with a and contains in C-to-N order the residues preceding a in the chain which are spatially close to a (distance below $dmax$).

2. Run Pratt twice, once on the complete set of C and once on the N strings. Search for patterns whose matches start from the beginning of the matched residue strings and that match residue strings from at least the minimum number of structures chosen by the user.

3. For each pattern, consider the neighbourhood string matches and retrieve the substructures (list of residues) corresponding to each such match. Compare the spatial geometry of the substructures by calculating the root mean square deviation (RMSD) when superposing each pair of substructures.

4. Output the patterns for which all pairwise RMSD values are below some upper limit, and rank them by I/R where I is the information content of the neighbourhood string pattern and R is the maximum RMSD for any pair of matching substructures.

The method differs from other structure comparison methods in that it considers information from all structures simultaneously in step 2. Other methods perform a number of pairwise structure comparisons and combine the results to find motifs shared by all or most of the structures under study. By utilizing information from all structures simultaneously, SPratt avoids considering patterns common to pairs of structures but not shared by the others. On the other hand, when Pratt is used to analyse the neighbourhood strings, many patterns can be found which later prove not to reflect similar substructures (i.e. they do not superpose very well). It might be advantageous to use additional structural information in this step to avoid considering such patterns. We explore extending the encoding of the neighbourhoods to include information about secondary structure and restrict Pratt to match only residues from the same sort of secondary structure (alpha-helix, beta-strand, or loop).

The patterns found by SPratt are evaluated using a very simple function, which basically rewards patterns containing more residues and imposing stricter restrictions on the amino acids allowed for each residue, and penalizes patterns for which the occurrences do not superpose very well using a measure of RMSD. However, the RMSD value for superposing the coordinates of a small number of residues, for example 4 is not very informative. The significance of the patterns found by Pratt, can be further assessed using the structure alignment program SAP (42). We have done this by rewarding alignment of residues in agreement with the motif, and in this way checking whether the matching of the few residues described by the motif can be extended to an alignment of larger parts of the structures.

6 Examples

In (15) we described the application of the first version of Pratt to the analysis of some protein families in PROSITE. For example, we analysed the Snake toxin family (PS00272 in PROSITE) containing 164 sequences of average length 64. We retrieved the sequences from the SWISS-PROT database and input them (un-aligned) to the Pratt program. Using default parameters (which requires patterns to match all the sequences), no patterns were found. However, using Pratt to discover patterns matching at least 155 out of the 164 sequences, we got the pattern G-C-x(1,3)-C-P-x(8,10)-C-C-x(2)-[PDEN]. This pattern turned out not to match any sequences in SWISS-PROT apart from the family members and was since included in the PROSITE database as the pattern for this family.

Using the SPratt program we analysed a set of cupredoxin protein structures (40). The proteins were selected from the cupredoxins super-family in SCOP (43) so that all pairwise sequence similarities were 30% or less. The 10 structures were input to the SPratt program and it was instructed to search for patterns containing single residue elements and also allowing the match-set [MLQ] (since the methionine ligand-binding residue is known to be substituted by L or Q in some proteins). SPratt used two minutes and identified three patterns all match-ing around the copper binding sites. The substructure occurrences of the pattern superpose with very low RMSD values (0.7 Å or less). We also used the motif identified by SPratt to guide the structure alignment program SAP using the pairs of equivalenced residues as extra constraints on the alignment. This resulted in a greatly improved alignment with RMSD of 1.56 Å over 63 pairs of residues (as compared to the alignment with RMSD of 5.1 Å over 26 residues when no SAP was run without any motif information).

7 Conclusions

Patterns and pattern discovery tools can help in the analysis of protein families, i.e. sets of proteins believed to share structural and/or functional properties. The most well conserved parts of the sequences or structures can be identified and it can be analysed whether the conserved patterns are statistically significant and therefore likely to have biological importance. Furthermore, the patterns can be used to identify additional related proteins. Patterns can describe protein properties at sequence level (sequence patterns) or at structure level (structure patterns). There exist a number of databases of protein families that give for each family one or several sequence patterns that can be used to identify more family members. We described in some detail the PROSITE, BLOCKS, Identify, PRINTS, and the Pfam databases. Structure pattern databases are starting to appear, for instance PROCAT is a database of 3-D enzyme active site templates (35).

For sequence patterns we made a distinction between deterministic and probabilistic pattern. Regular expression type patterns fall into the first class and profiles and HMMs fall into the second class. The different types of patterns used each have their strengths and weaknesses. For example, regular expression patterns are easily interpreted, but provide less expressive power than profiles or HMMs.

A number of methods have been developed for the automatic discovery of conserved patterns in protein sequences. We summarized a framework comprising a three-step approach that can be used to better understand the myriad of different methods. One sequence motif discovery program, Pratt, was described in some more detail, focusing on the more practical aspects of how it can be used to find conserved regular expression type patterns in unaligned protein sequences.

Structure motifs can describe properties at different levels, e.g. at atom group or residue level or at secondary structure level. Most structure comparison methods are for comparing pairs of structures, but some of these have been extended to the comparison of several by combining the results of a set of pairwise comparisons. We described in some detail a tool SPratt (Structure Pratt) which is able to find patterns of locally packed residues whose spatial positions are similar in a set of protein structures. This method utilizes information from all structures simultaneously in the search for conserved motifs.

Motifs provide a powerful classification tool for determining structure and function of proteins coming out of the numerous genome projects. Also, methods for finding conserved patterns in structures and sequences can be used to extract information from the large amounts of biomolecular data. In this chapter we have only considered patterns and motifs in proteins, but pattern discovery methods can also be used in other applications. For example, recently, a pattern discovery method was used to find putative regulatory elements in sets of gene upstream regions from the Yeast genome (44).

References

1. Bairoch, A., Bucher, P., and Hofman, K. (1996). *Nucleic Acids Res.*, **24**, 189.
2. Dayhoff, M. O., Schwartz, R. M., and Orcutt, B. C. (1978). In *Atlas of protein sequence and structure* (ed. M. O. Dayhoff), Vol. 5, Suppl. 3, p. 345. National Biomedical Research Foundation, Washington DC.
3. Henikoff, S. and Henikoff, J. G. (1992). *Proc. Natl. Acad. Sci. USA*, **89**, 10915.
4. Needleman, S. B. and Wunsch, C. D. (1970). *J. Mol. Biol.*, **48**, 443.
5. Altschul, S. F., Gish, W., Miller, W., Myers, E. W., and Lipman, D. J. (1990). *J. Mol. Biol.*, **215**, 403.

6. Lipman, D. J. and Pearson, W. R. (1985). *Science*, **277**, 1435.

7. Brazma, A., Jonassen, I., Eidhammer, I., and Gilbert, D. (1998). *J. Comp. Biol.*, **5**, 279.

8. Pietrokovski, S., Henikoff, J. G., and Henikoff, S. (1996). *Nucleic Acids Res.*, **24**, 197.

9. Attwood, T. K., Beck, M. E., Bleasby, A. J., Debtyarenko, K., and Smith, D. J. P. (1996). *Nucleic Acids Res.*, **24**, 182.

10. Gribskov, M., McLachland, A. D., and Eisenberg, D. (1987). *Proc. Natl. Acad. Sci. USA*, **84**, 4355.

11. Krogh, A., Brown, M. Miah, I. S., Sjoelander, K., and Haussler, D. (1994). *J. Mol. Biol.*, **235**, 1501.

12. Brown, M., Hughey, R., Krogh, A., Mian, I. S., Sjoelander, K., and Haussler, D. (1993). In *Proc. 1st Int. Conf. On Intell. Systems for Mol. Biol.*, pp. 47–53. AAAI Press.

13. Altschul, S. F., Carroll, R. J., and Lipman, D. J. (1989). *J. Mol. Biol.*, **207**, 309.

14. Neuwald, A. F. and Green, P. (1994). *J. Mol. Biol.*, **239**, 698.

15. Jonassen, I., Collins, J. F., and Higgins, D. G. (1995). *Protein Sci.*, **4**, 1587.

16. Rissanen, J. (1978). *Automatica-J.IFAC*, **14**, 465.

17. Brazma, A., Jonassen, I., Ukkonen, E., and Vilo, J. (1996). In *Proc. 4th Int. Conf. On Intell. Systems for Mol. Biol.*, pp. 34–43. AAAI Press.

18. Jonassen, I., Helgesen, C., and Higgins, D. (1996). Reports in Informatics, report no. 116. Dept. of Informatics, Univ. of Bergen, Norway.

19. Sternberg, M. J. E. (1991). *Nature*, **349**, 111.

20. Bairoch, A. and Boeckmann, P. (1992). *Nucleic Acids Res.*, **20**, 2019.

21. Lathrop, R., Webster, T., Smith, R., Winston, P., and Smith, T. (1993). In *Artificial intelligence and molecular biology* (ed. L. Hunter), pp. 211–58. AAAI Press/The MIT Press.

22. Tatusov, R. L., Altschul, S. F., and Koonin, E. V. (1994). *Proc. Natl. Acad. Sci. USA*, **91**, 12091.

23. Bucher, P. and Bairoch, A. (1994). In *Proc of 2nd Int. Conf. On Intell. Systems for Mol. Biol.*, pp. 53–61. AAAI Press.

24. Sonnhammer, E. L., Eddy, S. R., Birney, E., Bateman, A., and Durbin, R. (1996). *Nucleic Acids Res.*, **26**, 320.

25. Nevill-Manning, C. G., Wu, T. D., and Brutlag, D. L. (1998). *Proc. Natl. Acad. Sci. USA*, **95**, 5865.

26. Saqi, M. A. S. and Sayle, R. (1994). *Comput. Appl. Biosci.*, **10**, 545.

27. Sayle, R. A. and Milner-White, E. J. (1995). *Trends Biochem. Sci.*, **20**, 374.

28. Smith, H. O., Annau, T. M., and Chandrasegaran, S. (1990). *Proc. Natl. Acad. Sci. USA*, **87**, 826.

29. Smith, R. F. and Smith, T. F. (1990). *Proc. Natl. Acad. Sci. USA*, **87**, 118.

30. Smith, T. F. and Waterman, M. S. (1981). *J. Mol. Biol.*, **147**, 195.

31. Thompson, J. D., Higgins, D. G., and Gibson, T. J. (1994). *Nucleic Acids Res.*, **22**, 4673.

32. Taylor, W. R. (1988). *J. Mol. Evol.*, **28**, 161.

33. Jonassen, I. (1997). *Comput. Appl. Biosci.*, **13**, 509.

34. Wallace, A. C., Borkakoti, N., and Thornton, J. M. (1997). *Protein Sci.*, **6**, 2308.

35. http://www.biochem.ucl.ac.uk/bsm/PROCAT/PROCAT.html

36. Brown, N. P., Orengo, C. A., and Taylor, W. R. (1996). *Comput. Chem.*, **20**, 359.

37. Taylor, W. R. and Orengo, C. A. (1989). *J. Mol. Biol.*, **208**, 1.

38. Artymiuk, P. J., Porrette, A. R., Grindley, H. M., Rice, D. W., and Willett, P. (1994). *J. Mol. Biol.*, **243**, 327.

39. Nussinov, R. and Wolfson, H. J. (1991). *Proc. Natl. Acad. Sci. USA*, **88**, 10495.

40. Jonassen, I., Eidhmmer, I., and Taylor, W. R. (1999). *Proteins: Struct. Funct. Genet.*, **34**, 206.

41. Karlin, S. and Zhu, Z.-Y. (1996). *Proc. Natl. Acad. Sci. USA*, **93**, 8344.

42. Taylor, W. R. (1999). *Prot. Sci.*, **8**, 654

43. Murzin, A. G., Brenner, S. E., Hubbard, T., and Chothia, C. (1995). *J. Mol. Biol.*, **247**, 536.

44. Brazma, A., Jonassen, I., Vilo, J., and Ukkonen, E. (1998). *Genome Res.*, **8**, 1202.

Chapter 8

Comparison of protein sequences and practical database searching

Golan Yona and Steven E. Brenner*

Department of Structural Biology, Stanford University, USA.

*Department of Plant and Microbial Biology, University of California, Berkeley, USA

1 Introduction

During the last three decades a considerable effort has been made to develop algorithms that compare sequences of biological macromolecules (proteins, DNA). The purpose of such algorithms is to detect evolutionary, and thus structural and functional, relations among sequences. Successful sequence comparison would allow us to infer the biological properties of new sequences from data accumulated on related genes. For example, a similarity between a translated nucleotide sequence and a known protein sequence suggests a homologous coding region in the corresponding nucleotide sequence. Significant sequence similarity among proteins may imply that the proteins share the same secondary and tertiary structures, and have close biological functions. The prediction of unknown protein structures is often based on the study of known structures of homologous proteins.

Today, the routine procedure for analysis of a new protein sequence almost always starts with a comparison of the sequence to hand with the sequences in one or more of the main sequence databases. A new sequence is analysed by extrapolating the properties of its 'Neighbours' in a database search. Such methods have been applied during the last three decades with much success and have helped to identify the biological function of many protein sequences, as well as to reveal many distant and interesting relationships between protein families. Actually, more sequences have been putatively characterized by database searches than by any other single technology.

Detecting homology may often help in determining the function of new proteins. By definition, homologous proteins have evolved from the same ancestor protein. The degree of sequence conservation varies among protein families. Yet, homologous proteins almost always have the same fold (1–3). Although the common evolutionary origin of two proteins is almost never directly observed, we can deduce homology, with a high statistical confidence, given that the sequence similarity is significant.

In principle, similarity does not necessarily imply homology (similarity may be quantified whereas homology is a relation that either holds or does not hold). Therefore, similarity should be used carefully in attempting to deduce homology. The deduction of biological function out of sequence similarity is not straightforward, and sequence comparison procedures may lead to false conclusions when applied simple-mindedly. Today sequence comparison algorithms are accompanied with statistical estimates which provide a measure of statistical significance of the observed sequence similarities. These estimates can further help in assessing the significance of the similarity, and in many cases can lead to deduction of homology. The confidence in the deduction clearly depends on the level of statistical significance. In this view, database searches should be treated as experiments analogous to wet-lab characterization. Their use deserves the same care both in the design of the experiment and in the interpretation of results.

Planning a good experiment requires understanding of the methods being applied. Fundamentally, database searches are a simple operation: a query sequence is aligned with each of the sequences (called targets) in a database. A score is computed from each alignment, and the query/target pairs with the best scores are then reported to the user. Statistics are used to help improve the ability to interpret these scores and distinguish true relations between proteins from chance similarities. A more detailed description of this process, the sequence-comparison algorithms, the scoring schemes, and the statistics of sequence alignments is given next.

2 Alignment of sequences

During evolution, sequences have changed by insertions, deletions, and mutations. These evolutionary events may be traced today by applying algorithms for sequence alignment. Suppose that a DNA sequence **a** has evolved to the sequence **b** through substitutions, insertions and deletions. This transformation can be represented by an alignment where **a** is written above **b** with the common (conserved) bases aligned appropriately. For example, say that **a** = ACTTGA and **b** is obtained by substituting the second base from C to G, inserting an A between the second and the third bases, and by deleting the fifth base (G). The corresponding alignment will be:

$$
\begin{array}{lccccccc}
\mathbf{a} & = A & C & - & T & T & G & A \\
\mathbf{b} & = A & G & A & T & T & - & A \\
score & = 1 & 0 & -1 & 1 & 1 & -1 & 1
\end{array}
$$

We usually do not actually know which sequence evolved from the other. Therefore the events are not directional and insertion of A in **b** might have been a deletion of A in **a**.

In a typical application we are given two related sequences and we wish to

168

recover the evolutionary events that transformed one to the other. The goal of sequence alignment is to find the *correct* alignment that encodes the true series of evolutionary events that have occurred. The alignment can be assigned a score which accounts for the number of identities (a match of two identical letters), the number of substitutions (a match of two different letters), and the number of gaps (insertions/deletions). For example, in the alignment above, a score of 1 was given for each identity, a score of 0 was given for each sub-stitution, and a negative score of –1 was given for each gap. Overall, the align-ment scored 2, which is the sum of all pair scores and gap scores. In general, the scores for identities and substitutions which are used to score the alignment are called the **scoring matrix**, and the scores for gaps are called **gap penalties**. Altogether they are called the **scoring scheme** (see *Section 4.5* for details). With high (positive) scores for identities, and low (or negative) scores for substitutions and gaps, the basic strategy towards tracing the correct alignment seeks the alignment which scores best. In the following sections we describe in detail the common algorithms for sequence comparison. The discussion focuses on the comparison of protein sequences, but it holds for DNA sequences as well.

2.1 Rigorous alignment algorithms

There are several different alignment algorithms which have become a standard tool for biologists. The rigorous algorithms use dynamic programming to find the optimal alignment.

2.1.1 Global alignment

The first to propose a dynamic programming algorithm for comparison of macromolecules, were Needleman and Wunsch (4). Their algorithm performs a **global alignment** of the sequences; i.e. an alignment where all letters of **a** and **b** are accounted for. This type of alignment is appropriate when similarity is expected along the whole or most of the sequence.

Formally, let $s(a_i,b_j)$ be the similarity score of a_i, b_j (the scoring matrix) and let $\alpha > 0$ be the penalty for deleting/inserting of one amino acid. The score of an alignment with N_{ij} matches of a_i and b_j and N_{gap} insertions/deletions is defined as:

$$\sum_{i,j} N_{ij} \cdot s(a_i,b_j) - N_{gap} \cdot \alpha$$

In sequence evolution, an insertion or deletion of a segment (several adjacent amino acids) usually occurs as a single event. That is, the opening of the gap is the significant event. Therefore, most computational models assign a penalty for a gap of length k that is smaller than the sum of k independent gaps of length 1, by charging large penalty for opening a gap, and a smaller penalty for each extension (**affine** or linear gap penalty). If the penalty for gap of length k is $\alpha(k)$, and N_{k-gap} is the number of gaps of length k in a given alignment, then the score of this alignment is defined in this case as:

$$\sum_{i,j} N_{ij} \cdot s(a_i,b_j) - \sum_{k} N_{k-gap} \cdot \alpha(k)$$

169

The **global similarity** of sequences **a** and **b** is defined as the largest score of any alignment of sequences **a** and **b**, i.e.

$$S(\mathbf{a},\mathbf{b}) = max_{alignments} \{\sum_{i,j} N_{ij} \cdot s(a_i,b_j) - \sum_{k} N_{k\text{-}gap} \cdot \alpha(k)\}$$

In principle, the number of possible alignments is exponentially large, what makes it impossible to perform a direct search. However, a dynamic programming algorithm makes it possible to find the optimal alignment without checking all possible alignments, but only a very small portion of the search space. In brief, the idea is that every subalignment in the optimal alignment should be optimal as well (otherwise it would be possible to improve the overall alignment by improving the subalignments, in contrast with its definition as optimal alignment). Since any optimal subalignment (say, of the substring $a_1 a_2 \ldots a_i$ with the substring $b_1 b_2 \ldots b_j$) can end only in one of following three ways:

$$
\begin{array}{ccccc}
a_i & & a_i & & - \\
b_j & \text{or} & - & \text{or} & b_j
\end{array}
$$

every possible subalignment is calculated only once, and in constant time[1], out of its optimal subalignments.

Formally speaking, denote by $S_{i,j}$ the score of the best alignment of the substring $a_1 a_2 \ldots a_i$ with the substring $b_1 b_2 \ldots b_j$, i.e.

$$S_{i,j} = S(a_1 a_2 \ldots a_i, b_1 b_2 \ldots b_j)$$

Assume that the gap penalty is constant and equals α. Then, after an initialization step

$$S_{0,0} = 0 \qquad S_{i,0} = -i \cdot \alpha \quad \text{for } i = 1 \ldots n \qquad S_{0,j} = -j \cdot \alpha \quad \text{for } j = 1 \ldots m$$

(where n and m are the lengths of the sequences **a** and **b** respectively) define $S_{i,j}$ recursively

$$S_{i,j} = max\{S_{i-1,j-1} + s(a_1, b_j), S_{i,j-1} - \alpha, S_{i-1,j} - \alpha\}$$

Therefore, the score $S(\mathbf{a},\mathbf{b})$ can be calculated recursively. Since the subalignment for each i and j has to be calculated, the time complexity of this algorithm is proportional to the product of the lengths of the sequences compared (a quadratic time complexity). In practice, the scores are stored in a two-dimensional array of size $(n+1) \cdot (m+1)$. The initialization set the values at row zero and column zero and the computation proceeds row by row so that the value of each matrix cell is calculated from entries which were already calculated (see *Figure 1*).

2.1.2 Local alignment

In many cases the similarity of two sequences is limited to a specific motif or domain, the detection of which may yield valuable structural and functional insights, while outside of this motif/domain the sequences may be essentially

---

[1] This is true with linear gap functions. With non-linear gap penalties, the calculation of this optimal subalignment may need up to i+j+1 operations.

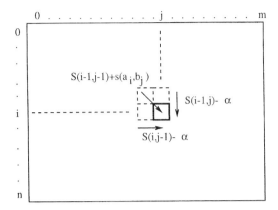

Figure 1 Calculating the global similarity score. The score of the (*i,j*) entry in the matrix is calculated from three matrix cells: the one on the left, the one on the top, and the one located at the top left corner of the current cell. In case of a non-constant gap penalty we need also to check all the cells in the same row and all the cells in the same column (along the dashed lines).

unrelated. In such cases global alignment may not be the appropriate tool. In the search for an optimal global alignment, local similarities may be masked by long unrelated regions. Consequently, the score of such an alignment can be as low as for totally unrelated sequences. Moreover, the algorithm may even mis-align the common region. Therefore, usually it is better to compare sequences locally. A **local alignment** of **a** and **b** is defined as an alignment between a *substring* of **a** and a *substring* of **b**. The **local similarity** of sequences **a** and **b** is defined as the maximal score over all possible local alignments.

The algorithm which finds the best local alignment is based on a minor mod-ification of the dynamic programming algorithm for global alignment. Specific-ally, whenever the score of the optimal subalignment of two subsequences becomes negative, the score is set to zero, meaning that the corresponding subsequences should not be aligned. Following the notations of the previous section, $S_{i,j}$ is now defined

$$S_{i,j} = max\{0, S_{i-1,j-1} + s(a_1, b_j), S_{i,j-1} - \alpha, S_{i-1,j} - \alpha\}$$

In the literature, this algorithm is often called the Smith–Waterman (SW) algorithm, after those who introduced this modification (5).

There is a lot of literature regarding dynamic programming algorithms in general (6), and for sequence comparison specifically (7–9). The interested reader is referred to these books for more details on the algorithmic aspects of this method, as well as its computational aspects.

2.2 Heuristic algorithms for sequence comparison

In a typical application new protein sequence is compared with all sequences in the database (**library sequences**), in search of related proteins. Because of its quadratic time complexity, the dynamic programming algorithms may not be suitable for this purpose. For example, the comparison of a sequence, of average length of 350 amino acids, against a typical database (like SWISSPROT (10), with more than 80 000 sequences), may take few CPU hours on a standard PC of nowadays (Pentium-III).

Several algorithms have been developed to speed up the alignment procedure. The two main algorithms are FASTA (11) and BLAST (12). These are heuristic algorithms which are not guaranteed to find the optimal alignment. However, they proved to be very effective for sequence comparison, and they are significantly faster than the rigorous dynamic programming algorithm.[2]

2.2.1 BLAST (Basic Local Alignment Search Tool)

BLAST compares two sequences and seeks all pairs of similar segments, whose similarity score exceeds a certain threshold. These pairs of segments are called 'high scoring segment pairs' (HSPs). A segment is always a contiguous sub-sequence of one of the two sequences. Segment pair is a pair of segments of the same length, one from each sequence. Hence the alignment of the segments is without gaps. The score of the match is simply the sum of matches of the amino acids (defined by a scoring matrix) along the segment pair. The segment pair with the highest score is called the 'maximum segment pair' (MSP).

To identify the HSPs (and particularly, the MSP), the algorithm starts by locating 'seeds' of similarity among the query sequence and the database sequence that score at least T, and then extends them in both directions until the maximum possible score for the extension is reached. The changes in the threshold T permit a tradeoff between speed and sensitivity. A higher value of T yields greater speed, but also an increased probability of missing weak similarities. Finally, multiple MSP regions are combined. For each consistent combination, its probability is calculated using the Poisson or sum statistics (14) and the most significant hits (lowest probability) are reported.

The algorithm is an outgrowth of the statistical theory for local alignments without gaps (see *Section 3*). This theory gives a framework for assessing the probability that a given similarity between two protein sequences (i.e. the MSP) could have emerged by chance. If the probability is very low, then the similarity is statistically significant and the algorithm reports the similarity along with its statistical significance. Though the algorithm may miss complex similarities which include gaps, the statistical theory of alignments without gaps provided a reliable and efficient way of distinguishing true homologies from

---

[2] In the last few years, biotechnology companies such as Compugen and Paracel, have developed special purpose hardware that accelerates the dynamic programming algorithm (13). This special-purpose hardware has again made the dynamic programming algorithm competitive with FASTA and BLAST, both in speed and in simplicity of use. However, meanwhile, FASTA and BLAST have become standard in this field and are being used extensively by biologists all over the world. Both algorithms are fast, effective, and do not require the purchase of additional hardware. BLAST has an additional advantage, as it may reveal similarities which are missed by the dynamic programming algorithm, for example when two similar regions are separated by a long dissimilar region.

chance similarities, thus making this algorithm an important tool for molecular biologists.

Current improvements of BLAST allow gapped alignments, by using dynamic programming to extend a central seed in both directions (15). This is complemented by PSI-BLAST, an iterative version of BLAST, with a position-specific score matrix (see *Section 4.5*) that is generated from significant alignments found in round i and used in round $i+1$. The latter may better detect weak similarities that are missed in database searches with a simple sequence query.

2.2.2 FASTA

FASTA is another heuristic that performs a fast sequence comparison. The algorithm starts by creating a hash table of all k-tuples (a string of length k) in the query sequence (usually, k = 1 or 2 for protein sequences, where k = 1 gives higher sensitivity). This table stores the k-tuples in a way which enables fast accession, and restoration of each k-tuple. Then, when scanning a library sequence, each k-tuple of the library sequence is looked up in the hash table, and if it is found (this means k-tuple identity) it is marked. At a second stage, the ten regions with the highest density of identities are rescanned. Common k-tuples which are on the same diagonal (same offset in both sequences), and not very far apart (the exact parameters are set heuristically), are joined to form a region (a gapless local alignment, or HSP in BLAST terminology). The regions are scored to account for the matches as well as the mismatches, and the best region is reported (its score is termed 'initial score' or 'init1'). Then, the algorithm tries to join nearby high scoring regions, even if they are not on the same diagonal (the corresponding score being termed 'initn score'). Finally, a bounded dynamic programming is run in a band around the best region, to obtain the 'optimized score'. If the sequences are related then the optimized score is usually much higher than the initial score.

3 Probability and statistics of sequence alignments

In the evolution of protein sequences, not all regions mutate at the same rate. Regions which are essential for the structure and function of proteins, are more conserved. Therefore, significant sequence similarity of two proteins may reflect a close biological function or a common evolutionary origin. The algorithms that were described in the previous section can be used to identify such similarities. However, on any two input protein sequences, even if totally unrelated, the algorithms almost always find some similarity. For unrelated sequences this similarity is essentially random. As the length of the sequences compared increase, this random similarity may increase as well. Therefore, in order to assess the significance of a similarity score it is important to know what score to expect simply by chance.

Naturally, we would like to identify those similarities which are genuine, and biologically meaningful. In the view of the last paragraph, the raw similarity score may not be appropriate for this purpose. However, when the sequence similarity is statistically significant we can deduce, with high confidence level, that the sequences are related.[3] The reverse implication is not always true. We encounter many examples of low sequence similarity despite functional and structural similarity (16–18).

Though statistically significant similarity is neither necessary nor sufficient for a biological relationship, it may give us a good indication of such relationship. When comparing a new sequence against the database, in search of close relatives, this is extremely important, as we are interested in reporting only significant hits, and sorting the results according to statistical significance seems reasonable.

To estimate the statistical significance of similarity scores, a statistical theory should be developed. A great effort was made in the last two decades to establish such statistical theory. Currently, there is no complete theory, though some important results were obtained. These results have very practical implications and are very useful for estimating the statistical significance of similarity scores. The statistical significance of similarity scores for 'real' sequences is defined by the probability that the same score would have been obtained for random sequences. The statistical results concern the similarity scores of random sequences, when the similarity scores are defined by ungapped alignments. However, these results have created a framework for assessing the statistical significance of various similarity scores, including gapped sequence alignments, and recently, even structural alignments (19).

Readers who are primarily interested in practical applications (rather than the statistical basis) of the methods might like to proceed to *Section 4*.

3.1 Statistics of global alignment

Though the distribution of global similarity scores of random sequences has not been characterized yet, some important properties of this distribution were partly determined. The main characteristic of this distribution is the linear growth (or decline, depends on the mean of the scoring matrix) with the sequence length, i.e. the **expected** global similarity score grows **linearly** with the sequence length. However, the growth rate has not been determined.

The statistical significance of a similarity score obtained for '**real**' sequences, which exceeds the expected score by a certain amount, is estimated by the probability that the similarity score of **random** sequences would exceed the expected mean by the same amount. However, since the distribution of scores is unknown, the available estimates give only a rough bound on that probability. The variance of the global similarity score has not been determined either, and the best results give only an upper bound.

In practice, it is possible to approximate the distribution empirically by shuffling the sequences and comparing the shuffled sequences. By repeating

---

[3] Two exceptions are segments with unusual amino acid composition, and similarity that is due to convergent evolution.

this procedure many times it is possible to estimate the mean and the variance of the distribution, and a reasonable measure of statistical significance (e.g. by means of the z-score) can be obtained. Formally, denote by S the global similarity score. Let μ and σ^2 be the mean and the variance of the distribution of scores. Then, the z-score associated with the score S is defined as

$$\frac{S - \mu}{\sigma}.$$

This score measures how many units of standard deviation apart the score S is from the mean of the distribution. The larger it is, the more significant is the score S.

3.2 Statistics of local alignment without gaps

The statistics of ungapped similarity scores has been studied extensively since the early 1990s. The exclusion of gaps allowed a rigorous mathematical treatment, and several important results were obtained. Karlin and Altschul (21) have shown that for two random sequences of length n and m, the score of the best ungapped local alignment (the MSP score in BLAST jargon) is centred around

$$\frac{\ln(n \cdot m)}{\lambda},$$

where λ is a parameter that depends on the overall **background distribution** of amino acids in the database, and the scoring matrix. That is, the local similarity score grows **logarithmically** with the length of the sequences, and with the size of the search space ($n \cdot m$).

This result in itself is still not enough to obtain a measure of statistical significance for local similarity scores. This can be done only once a concentration of measure result is obtained or the distribution of similarity scores is defined. Indeed, one of the most important results in this field is the characterization of the distribution of local similarity scores without gaps. This distribution was shown to follow the extreme value distribution (20–22).

Formally, as the **sum** of many random variables is distributed **normally**, then the **maximum** of many random variables is distributed as an **extreme value distribution** (23). This distribution is characterized by two parameters: the index value u and the decay constant λ (for $u = 0$ and $\lambda = 1$, the distribution is plotted in *Figure 2*). The distribution is not symmetrical. It is positive definite and unimodal with one peak at u. Practically, the score of the best local alignment (the MSP score) is the maximum of the scores of many independent alignments, which explains the observed distribution. Specifically, S, the local similarity score of two random sequences of length n and m, is distributed as an extreme value distribution and

$$Prob(S \geq x) \sim 1 - \exp(-e^{-\lambda(x-u)})$$

where $u = (\ln Kmn)/\lambda$, and K is a constant that can be estimated from the background distribution and the scoring matrix (Karlin & Altschul 1990).

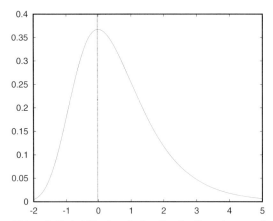

Figure 2 Probability density function for the extreme value distribution with u = 0 and λ = 1.

For a large x we can use the approximate $1 - \exp(-e^{-x}) \sim e^{-x}$. Therefore, for a large x,

$$Prob(S \geq x) \sim e^{-\lambda(x-u)} = e^{-\lambda \cdot x}e^{\lambda \cdot u} = Kmne^{-\lambda \cdot x}$$

This result helps to calculate the probability that a given MSP score could have been obtained by chance. The score will be statistically significant at the 1% level if $S \geq x_0$ where x_0 is determined by the equation $Kmne^{-\lambda \cdot x_0} = 0.01$. In general, a pairwise alignment with score S has a **p-value** of p where $p = Kmne^{-\lambda \cdot S}$. I.e., there is a probability p that this score could have happened by chance.

The probability p, that a similarity score S could have been obtained simply by chance from the comparison of two random sequences, should be adjusted when multiple comparisons are performed. One example of this is when a sequence is compared with each of the sequences in a database with D sequences. Denote by p-**match** a match between two sequences that has a p-value $\leq o$ (i.e. its score $\geq S$). The probability P of observing at least one p-match (i.e. at least one 'success'), in a database search follows the Poisson distribution

$$P = Prob(\text{at least one } p\text{-match}) = 1 - e^{-Dp}$$

and for $Dp < 0.1$ is well approximated by

$$P \simeq Dp$$

Since not all library sequences have the same probability of sharing a similar region with the query sequence, D should be replaced with the effective size of the database. If the query sequence is of length n, and the (pairwise) alignment of interest involves a libary segment of length m, and the database has a total of N amino acids, then D should be replaced with N/m. Thus,

$$P \simeq \frac{N}{m}p = KNne^{-\lambda \cdot S}$$

so the effective size of the search space is Nn (intuitively, this is the number of possible starting positions of a match).

It is very common to use the **expectation value** (e-value) as a measure of statistical significance. The expectation value of the Poisson distribution is given by

$$E = E(\text{number of } p\text{–matches}) = Dp$$

and as discussed above, D should be replaced with N/m. Hence

$$E = KNne^{-\lambda \cdot S}$$

This is the expected number of distinct matches (segment pairs) that would obtain a score $\geq S$ by chance in a database search, with a database of size N (amino acids) and composition **P** (the background distribution of amino acids). The higher it is, the match is less significant. For example, if $E = 0.01$, then the expected number of random hits with a score $\geq S$ is 0.01. In other words, we may expect a random hit with that score only once in 100 independent searches. If $E = 10$, then we should expect 10 hits with a score $\geq S$ by chance, in a single database search. This means that such a hit is not significant. (Note that $E \simeq P$ for $P < 0.1$.)

Finally, by setting a value for E and solving the equation above for S, it is possible to define a threshold score, above which hits are reported. This is the score above which the number of hits that are expected to occur at random is $< E$. Therefore, we can deduce that a match with this score or above reflects true biological relationship, but we should expect up to E errors per search. The specific value of E affects both the sensitivity of a search (the number of true relationships detected) and its selectivity (the number of errors). A lower value of E would decrease the error rate. However, it would decrease the sensitivity as well. A reasonable choice for E is between 0.1 and 0.001.

3.3 Statistics of local alignment with gaps

Though local alignments without gaps may detect most similarities between related proteins, and give a good estimation of the similarity of the two sequences, it is clear that gaps in local alignments are crucial in order to obtain the correct alignment, and for a more accurate measure of similarity. However, no precise model has been proposed yet to explain gaps in alignments. Moreover, introducing gaps in alignments greatly complicates their mathematical tractability. Rigorous results have been obtained only for local alignments without gaps.

Recent studies suggest that the score of local gapped alignments can be characterized in the same manner as the score of local ungapped alignments: As was mentioned in the previous section, the local **ungapped** similarity score grows logarithmically with the sequence's length and the size of the search space. Arratia and Waterman (24) have shown that for a range of substitution matrices and gap penalties, local **gapped** similarity scores have the same asymptotic characteristic. Furthermore, empirical studies (25, 26) strongly suggest that local gapped similarity scores are distributed according to the extreme value distribution, though some correction factors may apply (27). Based on empirical observations, Pearson (28) has derived statistical estimates for local alignment with gaps, using the extreme value distribution for scores obtained from a database search. A database search provides tens of thousands of scores from sequences which are unrelated to the query sequence, and therefore are effectively random. As discussed above, these scores are thus expected to follow the extreme-value distribution. This is true as long as the gap penalties are not too low. Otherwise the alignments shift from local to global and the extreme value distribution no longer apply.

Since the logarithmic growth in the sequence length holds in this case, scores are corrected first for the expected effect of sequence length. The correction is done by calculating the regression line $S = a + b \cdot \ln n$ for the scores obtained in a database search, after removing very high scoring sequences (probably related sequences). The process is repeated as many as five times. The regression line and the average variance of the normalized scores are used to define the **z-score**:

$$z\text{-score} = \frac{S - (a + b \cdot \ln n)}{var}$$

and the distribution of z-scores is approximated by the extreme value distribution

$$P = Prob(z\text{-score} > x) = 1 - \exp(-e^{c_1 \cdot x - c_2})$$

where c_1 and c_2 are constants, and the expectation value is defined as before by $E(z\text{-score} > x) = N \cdot p$ where N is the number of sequences in the database (the number of tests).

This empirical approach has the advantage of internal calibration of the accuracy of the estimates, and has proved to be very accurate in estimating the statistical significance of gapped similarity scores (28). (See also refs. 18 and 29.)

4 Practical database searching

4.1 Types of comparison

To formulate the database search 'experiment', it is first necessary to decide what types of sequences will be compared: DNA, Protein, or DNA as Protein. The algorithms described above may be applied to the comparison of protein sequences as well as to DNA sequences (coding or non-coding regions). However, the comparison of protein sequences has proven to be a much more effective tool (3). Though the evolutionary events occur at the DNA level, the main genetic pressure is on the protein sequence. Moreover, mutations at the DNA level do not necessarily change the encoded amino acid due to the redundancy of the genetic code. Mutations often result in conservative substitutions at the protein level, namely, replacement of an amino acid by another amino acid with similar biochemical properties. Such changes tend to have only a minor effect on the protein's functionality. Therefore, if the sequence under consideration either is a protein or codes for a protein, then it is almost always the case that

Table 1 Comparison programs and types of comparison

Programs	Query	DB	Comparison	Common Use
blastn, fasta, ssearch	DNA	DNA	DNA-level	Seek identical DNA sequences, and splicing patterns
blastp, fasta, ssearch	Protein	Protein	Protein-level	Seek homologous proteins
blastx, fastx	DNA	Protein	Protein-level	Query new DNA to find genes and seek homologous proteins
tblastn, tfasta, tfastx	Protein	DNA	Protein-level	Search for genes in un-annotated DNA
tblastx	DNA	DNA	Protein-level	Discover gene structure

the search should take place at the protein level, as proteins allow one to detect far more distant homologies than DNA. Another aspect is that in DNA comparisons, there is noise from comparisons of non-coding frames (though this latter issue still arises in DNA as Protein searches). DNA versus DNA comparison is typically only used to find identical regions of sequence in a database. One would do such a search to discover whether another group has sequenced or studied a gene, and to learn where it is expressed or where splice junctions occur. In short, protein-level searches are valuable for detecting evolutionarily related genes, while DNA searches are best for locating nearly identical regions of sequence (see *Table 1* for available comparison programs and the corresponding types of comparison).

4.2 Databases

Next, it is necessary to select a database to search against. There are several commonly used databases (e.g. GenBank, SwissProt, ESTs, etc.). For homology searches, it is best to use a comprehensive collection of all known proteins. Two such databases are available. One is the nr database at the NCBI website (http://www.ncbi.nlm.nih.gov/). The nr (which stands for non-redundant) protein database combines data from several sources (GenPept, SwissProt, PIR, RPF, and PDB) removes the redundant identical sequences, and yields a collection with nearly all known proteins. The second nr database is available at the ExPASy website in Switzerland (http://www.expasy.ch/). Both databases are frequently updated, to incorporate as many sequences as possible. Obviously, a search will not identify a sequence that has not been included in the database, and since databases are growing so rapidly, it is essential to use a current database.

The main sources of these non-redundant databases are the SwissProt database and the TrEMBL database (10), the PIR database (30), and the GenPept database (31). The SwissProt database is maintained at the ExPASy centre in Switzerland. This is a non-redundant highly annotated database which offers a lot of valuable biological information on almost all of its entries (more than 86 000 in the latest release, June 2000). Such information may include for example the description of the function of a protein, its domain structure, post-translational modifications, etc. This database is supplemented by TrEMBL, which is a collection of all the translations of EMBL nucleotide sequence entries not yet integrated in SwissProt. For most of these entries some biological information is available,

Table 2 Sequence databases. Number of entries is updated to June 2000.

Protein Database	Number Entries	Availability	Description
nr (ExPasy)	385002	ftp://www.expasy.ch/databases/sp_tr_nrdb/	consists of SwissProt, TrEMBL
nr (NCBI)	508388	ftp://ncbi.nlm.nih.gov/blast/db/	consists of GenPopt, SwissProt, PIR, RPF, PDB
SwissProt	86337	http://www.expasy.ch/sprot/sprot-top.html	non-redundant, high level of annotation
TrEmbl	298665	http://www.expasy.ch/sprot/sprot-top.html	non-redundant, computer annotated
PIR	180605	http://www-nbrf.georgetown.edu/pirwww/pirhome.shtml	non-redundant, annotated, family classification
GenPept	544510	http://www.ncbi.nlm.nih.gov/Entrez/protein.html	translation of DNA sequences in GenBank
PDB	12426	http://www.rcsb.org/pdb/index.html	repository of all known 3D structures
Genomes		http://www.ncbi.nlm.nih.gov/Entrez/Genome/org.html	protein sequences sorted by organism

DNA Database	Number Entries	Availability	Description
GenBank	5691170	http://www.ncbi.nlm.nih.gov/Entrez/nucleotide.html	DNA sequence
EMBL	5865742	http://www.ebi.ac.uk.embl/	DNA sequence
DDBJ	5962608	http://www.ddbj.nig.ac.jp/	DNA sequence

usually based on sequence analysis carried by the ExPASy team. PIR is another database that offers a lot of biological information on entries through an extensive annotation as well as classification to families and superfamilies and links to alignments with other family members. GenPept is a database that contains all translations of DNA sequences in the GenBank database.

Several specialized databases are also available, all of which overlap with the composite non-redundant databases. For example, if one is interested in searching for proteins of known structure, it is best to just search the smaller PDB database. Other specialized databases are available for each of the fully sequenced genomes, as well as for subsets of protein families (such as protein kinases or immunoglobulins), etc. See *Table 2* for a list of the main databases (see also Chapters 9 and 10).

One may also wish to search DNA databases at the protein level. Programs can do so automatically by first translating the DNA in all six reading frames and then making comparisons with each of these conceptual translations. The nr DNA database (containing most known DNA sequence except GSS, EST, STS, or HTGS sequences) is useful to search when hunting new genes; the identified genes in this database would already be in the protein nr database. Searches against the GSS, EST, STS, and HTGS databases can find new homologous genes, and are especially useful to learn about expression data or genome map location.

Table 3 Availability of sequence comparison programs

Program	FTP site	Run over the Web
ssearch	ftp://ftp.virginia.edu/pub/fasta/	http://www2.ebi.ac.uk/bic_sw/
		http://genome.dkfz-heidelberg.de/genweb/
		http://sgbcd.weizmann.ac.il/genweb/
		http://www.ch.embnet.org/software/FDF_form.html
fasta	ftp://ftp.virginia.edu/pub/fasta/	http://www2.ebi.ac.uk/fasta3/
blastp	ftp://ncbi.nlm.nih.gov/blast/	http://www.ncbi.nlm.nih.gov/BLAST/
		http://www2.ebi.ac.uk/blastall/
		http://www.ch.embnet.org/software/BottomBLAST.html

4.3 Algorithms

The choice of the comparison algorithm should be based on the desired comparison type, the available computational resources, and the goals of the search. All standard comparison algorithms can be run over the Web and can be downloaded from the FTP site to run locally (see *Table 3*). The rigorous Smith–Waterman algorithm is available, as well as the FASTA program, within the FASTA package. This algorithm is more sensitive than the others, but it is also much slower. The FASTA program is faster, and with the parameter *ktup* set to 1, is almost as sensitive as the Smith–Waterman algorithm (32, 18). The fastest algorithm is BLAST, the newest versions of which support gapped alignments (15) and provide a reliable, sensitive and fast option (the older versions are slower, detect fewer homologs, and have problems with some statistics). Iterative programs like PSI-BLAST require extreme care in their operation, as they can provide very misleading results; however, they have the potential to find more homologs than purely pairwise methods.

4.4 Filtering

The statistics for database searches assumes that unrelated sequences look essentially random with respect to each other. Specifically, the theoretical results that were obtained for the statistics of local alignments without gaps (see *Section 3.2*) are subject to the restriction that the amino acid composition of the two sequences that are compared are not too dissimilar (20). Assuming that both sequences are drawn from the background distribution, the amino acid composition of both should resemble the background distribution. Without this restriction the statistical estimates overestimates the probability of similarity scores, and indeed, this is observed in protein sequences with unusual compositions (18, 29). The most common exceptions are long runs of a small number of different residues (such as a poly-alanine tract). Such regions of a sequence may spuriously obtain extremely high match scores. For this reason, it is recommended to filter out these regions using programs such as SEG (33). The NCBI

181

BLAST server will automatically remove such sections in proteins, replacing them with X, if default filtering is selected. DNA sequences will be similarly masked by DUST. Though these programs automatically remove the majority of problematic matches, some problems invariably slip through; moreover, valid hits may be missed due to masking of part of the sequence. Therefore, it may be helpful to try using different masking parameters.

Other sorts of filtering are also often desirable; for example, iterative searches are prone to contamination by regions of proteins that resemble coiled-coils or transmembrane helices. Here, one protein that is similar only because it has the general characteristics may match initially. The profile then emphasizes these inappropriate characteristics, eventually causing many spurious hits. Heavily cysteine rich proteins can also obtain anomalous high scores. If these characteristics are not filtered, then it is necessary to carefully review the alignment results to ensure that they have not led to incorrect matches.

4.5 Scoring matrices and gap penalties

The next step is to choose the set of parameters for the sequence comparison algorithm. Namely, the scoring matrix and the gap parameters. The default matrices offered with the comparison algorithm (e.g. BLOSUM62 with BLAST, BLOSUM50 with FASTA) are a safe choice. However, it may be fruitful to check other matrices as well. Several different approaches were taken to derive reliable and effective scoring matrices. The most effective matrices are those that are based on actual frequencies of mutations that are observed in closely related proteins. These matrices reflect the biochemical properties of the amino acids, which influence the probability of mutual substitution (exchange occur more frequently among amino acids that share certain properties), and amino acids with similar properties have high pairwise score. Matrices which are based on sequence alignments include the family of PAM matrices (34) (and their improvement by ref. 35), the BLOSUM matrices (36), and Gonnet matrix (37). Other matrices, which proved to be very effective for protein sequence comparison, are those that are based on structural principles and aligned structures (38, 39).

The two most extensively used families of scoring matrices are the PAM matrices and the BLOSUM matrices. A detailed description of these matrices is given in the next two sections.

4.5.1 The PAM family of scoring matrices

PAM matrices were proposed by Dayhoff *et al.* in 1978 based on observations of hundreds of alignments of closely related proteins. The frequencies of substitution of each pair of amino acids were extracted from alignments of proteins of small evolutionary distance, below 1% divergence, i.e. at most one mutation per 100 amino acids, on average. These frequencies, normalized to account for the frequencies of random occurrences of single amino acids, resulted in the PAM-1 probability transition matrix. The PAM-1 matrix reflects an amount of evolutionary change that yields on average one mutation per 100 amino acids. Accordingly, it is suitable for comparison of proteins which have diverged by 1%

or less. The acronym PAM stands for Percent of Accepted Mutations (and hence the distance is in percentages) or for Point Accepted Mutations (and hence the distance in number of mutations per 100 amino acids).

The PAM-1 matrix is then extrapolated to yield the family of PAM-k matrices. Each PAM-k matrix is obtained from PAM-1 by k consecutive multiplication, and is suitable for comparison of sequences which have diverged k%, or are k evolutionary units apart. For example, PAM-250 = $(PAM-1)^{250}$ reflects the frequencies of mutations for proteins which have diverged 250% (250 mutations per 100 amino acids). The actual scoring matrices that are used by search programs are derived from the transition probability matrices and the background probabilities. The score of each pair $s(a,b)$ is defined as the logarithm of the likelihood ratio of the transition probability M_{ab} (mutation) versus the probability of a random occurrence of the amino acid b in the second sequence, i.e., $s(a,b) = \log M_{ab}/p_b$.

The PAM matrices were later refined by Jones $et\ al.$ (35) based on much larger data set. The significant differences were detected for substitions that were hardly observed in the original data set of (34).

The PAM-250 matrix. The PAM-250 matrix is one of the most extensively used matrices in this field. This matrix corresponds to a divergence of 250 mutations per 100 amino acids. Naturally one may ask whether it makes sense to compare sequences which have diverged this much. Surprising as it may seem, when calculating the probability that a sequence remains unchanged after 250 PAMs (this is given by the sum $\Sigma_a p_a M_{aa}$ where p_a is the probability of a random occurrence of amino acid s and M_{aa} is the diagonal entry in the PAM-250 matrix that corrresponds to the amino acid a) the outcome is that such sequences are expected to share about 20% of their amino acids. For reference, note that the expected percentage of identity in a random match is $100 \cdot \Sigma_a p_a^2$, and for a typical distribution of amino acids (in a large ensemble of protein sequences), we should expect less than 6% identies.

4.5.2 The BLOSUM family of scoring matrices

Unlike PAM matrices, which are extrapolated from a single matrix PAM-1, the BLOSUM series of matrices was constructed by direct observation of sequence alignments of related proteins, at different levels of sequence divergence. The matrices are based on 'blocks'—a collection of multiple alignments of similar segments without gaps (40), each block representing a conserved region of a protein family. These blocks provide a list of (accepted) substitutions, and a log-odds scoring matrix can be defined based on the observed relative frequency of aligned pairs of amino acids q_{ab}, and the expected probability of pairs e_{ab} estimated from the population of all observed pairs

$$s_{ab} = \log \frac{q_{ab}}{e_{ab}}$$

To reduce the bias in the amino acid pair frequencies caused by multiple counts from closely related sequences, segments in a block with at least x% identity are clustered and pairs are counted between clusters, i.e., pairs are counted only between segments less than $x\%$ identical. When counting pairs frequencies between clusters, the contributions of all segments within a cluster are averaged, so that each cluster is weighted as a single sequence. Varying the percentage of identity x within clusters results in a family of matrices BLOSUM-x, where x ranges from 30 to 100. For example, BLOSUM-62 is based on pairs that were counted only between segments less than 62% identical.

4.5.3 Choosing the scoring matrix

When comparing two sequences, the most effective matrix to use is the one which corresponds to the evolutionary distance between them (41). However, we usually do not know this distance. Therefore, it is recommended to use several scoring matrices which cover a range of evolutionary distances, for example PAM-40, PAM-120, and PAM-250. In general, low PAM matrices are well suited to finding short but strong similarities, while high PAM matrices are best for finding long regions of weak similarity.

Exhaustive evaluations have been carried out to compare the performance of different scoring matrices (42, 32). These studies show that log-odds matrices derived directly from alignments of highly conserved regions of proteins (such as BLOSUM matrices or the Overington matrix, which is based on structural alignment (39)) outperform extrapolated log-odds matrices based on an evolutionary model, such as PAM matrices. Moreover, the accuracy of alignments based on extrapolated matrices decreases as the evolutionary distance increases. This suggests that extrapolation cannot accurately model distant relationships, and that the PAM evolutionary model is inadequate. BLOSUM matrices were shown to be more effective in detecting homologous proteins. Specifically, BLOSUM-62 and BLOSUM-50 gave superior performance in detecting weak homologies. These matrices offer good overall performance in searching the databases. The best hybrid of matrices for searching in different evolutionary ranges is either BLOSUM 45/62/100 or BLOSUM 45/100 plus the Overington matrix.

4.5.4 Gap penalties

There is no mathematical model to explain the evolution of gaps. Practical considerations (the need for a simple mathematical model, time complexity) have led to the broad use of linear gap functions, where the penalty for a gap of length k is given by $\alpha(k) = \alpha_0 + k \cdot \alpha_1$. Usually a large penalty is charged for opening a gap (α_0), and a smaller penalty is charged for each extension (α_1).

Gonnet et al. (37) have proposed a model for gaps that is based on gaps occurring in pairwise alignments of related proteins. The model suggests an exponentially decreasing gap penalty function. However, a linear penalty function has the advantage of better time complexity, and in most cases the results are satisfactory. Therefore the use of linear gap functions is very common.

The gap parameters that are used as default in the standard comparison programs are usually optimized based on extensive evaluations (32), and it is rarely beneficial to change these from their defaults.

4.5.5 Position dependent scores

In many proteins, mutations are not equally probable along the sequence. Some regions are functionally/structurally important and consequently, the effect of mutation in these regions can be drastic. They may create a non-functional protein or even prevent the molecule from folding into its native structure. Such mutations are unlikely to survive, and therefore these regions tend to be more evolutionary conserved than other, less constrained regions (e.g. loops) which can significantly diverge.

Accordingly, it may be appropriate to use position-dependent scores for mismatches and gaps. The incorporation of information about structural preferences can lead to alignments that are more accurate biologically. If a protein's structure is known, the secondary structure should be taken into account. In the absence of such data, general structural criteria, such as the propensities of amino acid for occurring in secondary structures versus loops can be taken into account. For example, the probability of opening a gap in existing secondary structure can be decreased, while the probability for opening/inserting a gap in loop regions can be increased.

Usually position-specific scoring matrices, or **profiles**, are not tailored to a specific sequence. Rather, they are built to utilize the information in a group of related sequences, and provide representations of protein families and domains. These representations are capable of detecting subtle similarities between distantly related proteins. Without going into detail, profiles are usually obtained by applying algorithms for multiple alignment (i.e. a combined alignment of several proteins) to align a group of related sequences. The frequency of each amino acid at each position along the multiple alignment is then calculated. These counts are normalized and transformed to probabilities, so that a probability distribution over amino acids is associated with each position. Finally, the scoring matrix is defined based on these probability distributions as well as on the similarities of pairs of amino acids (taken from a standard scoring matrix). For example, the score for aligning the amino acid a at position i of the profile is given by

$$s_i(a) = \Sigma_b prob(b \text{ at position } i)s(a, b)$$

where $s(a,b)$ is the similarity of amino acids a and b according to some scoring matrix. For a review on algorithms for multiple alignment and profile techniques, see refs 7–9, 43, 44.

4.6 Command line parameters

The command line parameters of the search programs are generally divided into three groups. The first group is the set of parameters which specify the input and output filenames, and the database name. These are the only mandatory parameters. All other parameters are optional and are set default values otherwise. For example, the basic command line for SSEARCH, FASTA, BLAST, and gapped BLAST are:

ssearch -Q query-file -O out-file database
fasta -Q query-file -O out-file database
blastp database query-file
blastpgp -i query-file -o out-file -d database

The second set of parameters affects the comparison algorithm. This set includes the scoring matrix and the gap penalties and the parameters used to control the sensitivity of the search. By altering the later, it is possible to make the program run slower and be more sensitive, or to run faster at the cost of missing more homologs. BLAST has few such parameters. Currently, it is very rare for users to alter these options from the defaults. The FASTA program has one such parameter that a user will often want to set, called *ktup*. Searches with *ktup* = 1 are slower, but are more sensitive than BLAST;*ktup* = 2 is faster but less effective.

Table 4 Parameters for sequence comparison programs. PSI-BLAST and gapped BLAST are executed by the same program (blastpgp). The default mode is a simple gapped BLAST (i.e., the parameter j is set to 1).

Program	Parameter	Use
search	-Q filename	query file
	-O filename	output file
	-E evalue	evalue threshold (only hits with evalue below this threshold are reported)
	-d number	maximal number of alignments displayed
	-H	suppresses histogram of scores
fasta	-Q filename	query file
	-O filename	output file
	-E evalue	evalue threshold (only hits with evalue below this threshold are reported)
	-d number	maximal number of alignments displayed
	-H	suppresses histogram of scores
	ktup number	controls sensitivity (can be either 1 or 2 for proteins and up to 4 for DNA)
blastp	E=evalue	evalue threshold (only hits with evalue below this threshold are reported)
	V=number	maximal number of hits reported
	B=number	maximal number of alignments displayed
	H=1	display histogram of scores
blastpgp	-d database	the database searched
	-i filename	query file
	-o filename	out file
	-e evalue	evalue threshold (only hits with evalue below this threshold are reported)
	-v number	maximal number of hits reported
	-b number	maximal number of alignments displayed
	-j number	maximal number of iterations (PSI-BLAST)
	-C filename	saves a checkpoint profile in a file after each iteration (PSI-BLAST)
	-R filename	reads the initial profile from a file (PSI-BLAST)
	-h evalue	evalue threshold for inclusion in a profile (PSI-BLAST)

Finally, there is a third set of parameters which controls the output of the program, e.g. how many results are reported, and how many alignments are displayed. The number of hits reported is often controlled by the e-value parameter (see *Section 3.2*). For example, by default, the BLAST programs will report only matches with an e-value up to 10 (this parameter also affects the sensitivity of the method, in an indirect manner). The total number of matches is limited to the best 500, and detailed information with the alignment is provided for up to 100 pairs. To retrieve more matches, these numbers can be altered (see *Table 4*).

5 Interpretation of results

Interpretation of the results of a sequence database search involves first evaluating the matches, to determine whether they are significant and therefore imply homology. The most effective way of doing so is through use of the statistical scores (the e-values). The e-values are more useful than the raw or bit scores, and they are far more powerful than percentage identity (which is best not even considered unless the identity is very high) (18). Fortunately, the e-values from FASTA, SSEARCH, and gapped BLAST seem to be accurate and are therefore easy to interpret (18, 29).

The e-value (or expectation-value) of a match should measure the expected number of sequences in the database which would achieve a given score. Therefore, in the average database search, one expects to find ten random matches with e-value score of 10; obviously, such matches are not significant. However, lacking better matches, sequences with these scores may provide hints of function or suggest new experiments. Scores below 0.01 would occur by chance only very rarely, and are therefore likely to indicate homology, unless biased in some way. Scores of near 1e-50 are now seen frequently, and these offer extremely high confidence that the query protein is evolutionarily related to the matched target in the database.

Inferring function from the homologous matched sequences is a process still fraught with difficulty. If the score is extremely good and the alignment covers the whole of both proteins, then there is a good chance that they will share the same or a related function. However, is dangerous to place too much trust in the query having the same function as the matched protein: functions do diverge, and organismal or cellular roles may alter even when biochemical function is unchanged. Moreover, a significant fraction of functional annotations in databases are wrong (45), so one needs to be suspicious. There are other complexities; for example, if only a portion of the proteins align, they may share a domain which only contributes an aspect of the overall function. It is often the case that all of the highest-scoring hits align to one region of the query, and matches to other regions need to be sought much lower in the score ranking. For this reason, it is necessary to consider carefully the overlap between the query and each of the targets.

Database search methods are also limited because most homologous sequences have diverged too far to be detected by pairwise sequence comparison methods (16–18). Thus, failure to find a significant match does not necessarily indicate that no homologs exists in the database. In such cases more sophisticated methods must be applied. For example, iterative search programs such as the profile based PSI-BLAST program (15) or the HMM based SAM-T98 (46) are advanced and sensitive search tools. However, these programs should be carefully used as they can lead to false positives by diverging from the original query sequence, and creating a profile that represents unrelated sequences.

The most powerful tools today are those that incorporate information from a group of related sequences. This strategy has led to the compilation of databases of protein families and domains. These databases have become an important tool in the analysis of newly discovered protein sequences. They usually offer biologically valuable information about domains and the domain structure of proteins, through multiple alignments and schematic representations of proteins, and can help to detect weak relationships between remote homologs. Such methods are described in chapters 3, 4, and 5. However, family databases are limited because they typically contain less than half of the proteins in sequence databases. Moreover, many families have not yet been characterized, while others are currently too sparse to yield reliable models. For this reason, database searches are crucial in the analysis of newly sequenced genes that have no clear homologs with known families, and by integrating the information obtained from a database search one may discover clues about the function of the new gene.

6 Conclusion

One should neither have excessive faith in the results of a database search, nor should they be blithely disregarded. The standard search programs such as FASTA, gapped BLAST and SSEARCH are well-tested and reliable indicators of sequence similarity, and their underlying principles are straightforward. These programs and their parameters have been optimized for the hundreds of thousands of runs every day. If one is careful about posing the database search experiment and interprets the results with care, sequence comparison methods can be trusted to rapidly and easily provide an incomparable wealth of biological information.

References

1. Sander, C. and Schneider, R. (1991). Database of homology-derived protein structures and the structural meaning of sequence alignment. *Proteins*, **9**, 56.
2. Hilbert, M., Bohm, G., and Jaenicke, R. (1993). Structural relationships of homologous proteins as a fundamental principle in homology modeling. *Proteins*, **17**, 138.
3. Pearson, W. R. (1996). Effective protein sequence comparison. In *Methods in enzymology* (ed. R. F. Doolittle), Vol. 266, pp. 227–58. Academic Press.
4. Needleman, S. B. and Wunsch, C. D. (1970). A general method applicable to the search for similarities in the amino acid sequence of two proteins. *J. Mol. Biol.*, **48**, 443.

5. Smith, T. F. and Waterman, M. S. (1981). Comparison of Biosequences. *Adv. App. Math.*, **2**, 482.

6. Cormen, T. H., Leiserson, C. E., and Rivest, R. L. (1990). *Introduction to algorithms*. MIT Press/McGraw-Hill Book Company.

7. Waterman, M. S. (1995). *Introduction to computational biology*. Chapman & Hall, London.

8. Setubal, J. C. and Meidanis, J. (1996). *Introduction to computational molecular biology*. PWS Publishing Co., Boston.

9. Gusfield, D. (1997). *Algorithms on strings, trees, and sequences: computer science and computational biology*. Cambridge University Press, Cambridge.

10. Bairoch, A. and Apweiler, R. (1999). The SWISS-PROT protein sequence data bank and its supplement TrEMBL in 1999. *Nucleic Acids Res.*, **27**, 49.

11. Pearson, W. R. and Lipman, D. J. (1988). Improved tools for biological sequence comparison. *Proc. Natl. Acad. Sci. USA*, **85**, 2444.

12. Altschul, S. F., Carrol, R. J., and Lipman, D. J. (1990). Basic local alignment search tool. *J. Mol. Biol.*, **215**, 403.

13. Compugen LTD. BIOCCELERATOR Manual. http://www.compugen.co.il

14. Altschul, S. F., Boguski, M. S., Gish, W. G., and Wootton, J. C. (1994). Issues in searching molecular sequence databases. *Nature Genet.*, **6**, 119.

15. Altschul, S. F., Madden, T. L., Schaffer, A. A., Zhang, J., Zhang, Z., *et al.* (1997). Gapped BLAST and PSI-BLAST: a new generation of protein database search programs. *Nucleic Acids Res.*, **25**, 3389.

16. Murzin, A. G. (1993). OB(oligonucleotide/oligosaccharide binding)-fold: common structural and functional solution for non-homologous sequences. *EMBO J.*, **12:3**, 861.

17. Pearson, W. R. (1997). Identifying distantly related protein sequences. *Comp. Appl. Biosci.*, **13:4**, 325.

18. Brenner, S. E., Chothia, C., and Hubbard, T. J. P. (1998). Assessing sequence comparison methods with reliable structurally identified distant evolutionary relationships. *Proc. Natl. Acad. Sci. USA*, **95**, 6073.

19. Levitt, M. and Gerstein, M. (1998). A unified statistical framework for sequence comparison and structure comparison. *Proc. Natl. Acad. Sci. USA*, **95**, 5913.

20. Karlin, S. and Altschul, S. F. (1990). Methods for assessing the statistical significance of molecular sequence features by using general scoring schemes. *Proc. Natl Acad. Sci. USA*, **87**, 2264.

21. Dembo, A. and Karlin, S. (1991). Strong limit theorems of empirical functionals for large exceedances of partial sums of i.i.d variables. *Ann. Prob.*, **19**, 1737.

22. Dembo, A., Karlin, S., and Zeitouni, O. (1994). Limit distribution of maximal non-aligned two-sequence segmental score. *Ann. Prob.*, **22**, 2022.

23. Gumbel, E. J. (1958). *Statistics of extremes*. Columbia University Press, New York.

24. Arratia, R. and Waterman, M. S. (1994). A phase transition for the score in matching random sequences allowing depletions. *Ann. Appl. Prob.*, **4**, 200–225.

25. Smith, T. F., Waterman, M. S., and Burks, C. (1985). The statistical distribution of nucleic acid similarities. *Nucleic Acids Res.*, **13**, 645.

26. Waterman, M. S. and Vingron, M. (1994). Rapid and accurate estimates of statistical significance for sequence data base searches. *Proc. Natl. Acad. Sci. USA*, **91**, 4625.

27. Altschul, S. F. and Gish, W. (1996). Local alignment statistics. In *Methods in enzymology* (ed. R. F. Doolittle). Vol. 266, pp. 460–80. Academic Press.

28. Pearson, W. R. (1998). Empirical statistical estimates for sequence similarity searches. *J. Mol. Biol.*, **276**, 71.

29. Yona, G., Linial, N., and Linial, M. (1999). ProtoMap: Automatic classification of protein sequences, a hierarchy of protein families, and local maps of the protein space. *Proteins*, **37**, 360.

30. George, D. G., Barker, W. C., Mewes, H. W., Pfeiffer, F., and Tsugita, A. (1996). The PIR-International protein sequence database. *Nucleic Acids Res.*, **24**, 17.

31. Benson, D. A., Boguski, M. S., Lipman, D. J., Ostell, J., Ouellette, B. F., Rapp, B. A., *et al.* (1999). GenBank. *Nucleic Acids Res.*, **27**, 12.

32. Pearson, W. R. (1995). Comparison of methods for searching protein sequence databases. *Protein Sci.*, **4**, 1145

33. Wootton, J. C. and Federhen, S. (1993). Statistics of local complexity in amino acid sequences and sequence databases. *Comput. Chem.*, **17**, 149.

34. Dayhoff, M. O., Schwartz, R. M., and Orcutt, B. C. (1978). A model of evolutionary change in Proteins. In *Atlas of protein sequence and structure* (ed. M. Dayhoff), Vol. 5, Suppl. 3, pp 345–52. National biomedical research foundation, Silver Spring, MD.

35. Jones, D. T., Taylor, W. R., and Thornton, J. M. (1992). The rapid generation of mutation data matrices from protein sequences. *Comput. Appl. Biosci.*, **8:3**, 275.

36. Henikoff, S. and Henikoff, J. G. (1992). Amino acid substitution matrices from protein blocks. *Proc. Natl Acad. Sci. USA*, **89**, 10915.

37. Gonnet, G. H., Cohen, M. A., and Benner, S. A. (1992). Exhaustive matching of the entire protein sequence database. *Science*, **256**, 1443.

38. Risler, J. L., Delorme, M. O., Delacroix, H., and Henaut, A. (1988). Amino acid substitutions in structurally related proteins. A pattern recognition approach. Determination of a new and efficient scoring matrix. *J. Mol. Biol.*, **204**, 1019.

39. Johnson, M. S. and Overington, J. P. (1993). A structural basis for sequence comparisons. An evaluation of scoring methodologies. *J. Mol. Biol.*, **233**, 716.

40. Henikoff, S. and Henikoff, J. G. (1991). Automated assembly of protein blocks for database searching. *Nucleic Acids Res.*, **19**, 6565.

41. Altschul, S. F. (1991). Amino acid substitution matrices from an information theoretic perspective. *J. Mol. Biol.*, **219**, 555.

42. Henikoff, S. and Henikoff, J. G. (1993). Performance evaluation of amino acid substitution matrices. *Proteins*, **17**, 49.

43. Gribskov, M. and Veretnik, S. (1996). Identification of sequence patterns with profile analysis. In *Methods in enzymology* (ed. R. F. Doolittle). Vol. 266, pp. 198–211. Academic Press.

44. Taylor, W. R. (1996). Multiple protein sequence alignment: algorithms and gap insertion. In *Methods in enzymology* (ed. R. F. Doolittle). Vol. 266, 343–67. Academic Press.

45. Brenner, S. E. (1999). Errors in genome annotation. *Trends Genet.*, **15**, 132.

46. Karplus, K., Barrett, C., and Hughey, R. (1998). Hidden markov models for detecting remote protein homologies. *Bioinformatics*, **14:10**, 846.

Chapter 9
Networking for the biologist

R. A. Harper

EMBL-European Bioinformatics Institute, Wellcome Trust Genome Campus, Cambridge, UK.

> Riding along in my automobile my baby beside me at the wheel
> Cruising and playing the radio with no particular place to go

Chuck Berry

1 Introduction

Every research worker would like to have the tools on hand to make his job quicker and more efficient, and with the advent of the World Wide Web many of the tasks associated with molecular biology have become freely available online. In the past when a scientist wanted to know something about a particular subject then the first option was to talk to colleagues in the laboratory and ask for their advice. If that was not sufficient then it was off to the library to scan abstracts or the latest journals for the relevant information.

However times are changing and so are working habits. Why ask questions from people in your laboratory when you can ask the same question on the Bionet newsgroups *http://www.bio.net* from research workers all over the world? Why thumb through textbooks for references when you can type in keywords to an Internet search engine such as Lycos or Alta Vista and get a satisfactory answer in no time at all? But often you find that the major search engines index everything on the Web, which makes it difficult to find exactly what you want. So often it is more profitable to use search engines that are totally dedicated to biology.

In Europe you could use BiowURLd *http://search.ebi.ac.uk:8888/compass/* or Bio-Hunt *http://www.expasy.ch/BioHunt*, which deal exclusively with biology-related subjects. Another comprehensive listing exists at the Virtual Library in the BioSciences division. *http://www.vlib.org/Biosciences.html*, and from China there is the NEE-HOW project, *http://biology.neehow.org* which is an invaluable resource for research workers from the Pacific rim.

In the USA one of the original and best lists of Biological resources, put together by Keith Robison can be found at Harvard *http://golgi.harvard.edu/biopages.list* and of course there is the ever popular *Pedro's BioMolecular Research Tools* at

http://www.public.iastate.edu/~pedro/research_tools.html. If you are looking specifically for software related to bioinformatics, then there is the BioCatalogue at
http://www.ebi.ac.uk/biocat/biocat.html or if you are looking for an obscure database
then there is DBcat *http://www.infobiogen.fr/services/dbcat* from Infobiogen the
EMBnet node in France.

There are also a few good newsletters, which deal specifically with what is
happening in the world of bioinformatics. EMBnet produces a quarterly newsletter, which gives an update of the latest developments at the different EMBnet
nodes throughout Europe. The EMBnet News can be found at the URL
http://www.ebi.ac.uk/embnet.news/embnetmenu.html.

The EBI has its own industry programme and they produce a newsletter called
the Bioinformer: *http://bioinformer.ebi.ac.uk/newsletter/.* One special feature within
this newsletter is the BioEvents Calendar, *http://bioinformer.ebi.ac.uk/Events/* that
allows people to advertise workshops, conferences, or symposiums. In the USA
there are two major newsletters associated with bioinformatics. The NCBI newsletter is at *http://www.ncbi.nlm.nih.gov/Web/Newsltr/index.html* and the National
Centre for Genomic Research, the NCGR newsletter, is at *http://www.ncgr.org/
ncgr/ncgr_newsletter.html*

The focus of this article is to help research workers avoid the World Wide
Wait while using World Wide Web.

2 The changing face of networking

In the early 1990s academic research workers had the networks all to themselves. Today however, the demography of those using the networks and their
reasons for using the networks have completely changed. The competition for
bandwidth is fierce between the commercial and academic sector.

The Georgia Institute of Technology (*http://www.gvu.gatech.edu/user_surveys/*)
has been conducting user surveys on the use of the Internet since 1994 (see
Figure 1). Over a four-year period there have been many radical changes in
attitude towards the use and abuse of the Web. The most recent surveys show
that when it comes to using the WWW the two main activities that people
engage in are collecting personal information, and using the Web purely for
entertainment. The academic no doubt will be distressed that work and education only occupy equal third place. Academics are no longer the only people
using the internet and they may feel that their research work suffers because of
the 'info-tourists' on the web. Gone are the days when the only people on the
network were scientists with Unix boxes. More and more people are coming
online from home and the humble PC seems to have cornered the market (see
Figure 2).

In the past scientists relied on centralized systems, with systems administrators installing and maintaining programmes. Nowadays since the installation of
many programmes on PC's has been fully automated, scientists are doing it for
themselves. This means that the scientist needs to be aware of the trends that
are driving the internet forward. Applications will be written for platforms that

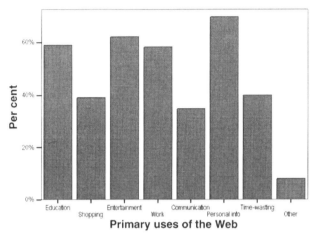

Figure 1 Primary uses of the Web. (Copyright 1994–1998 Georgia Tech Research Corporation. All rights Reserved. Source: GVU's WWW User Survey *www.gvu.gatech.edu/user_surveys*)

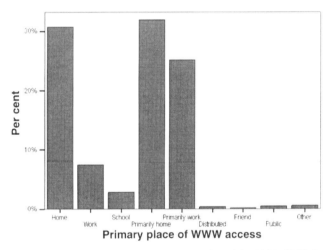

Figure 2 Primary places of WWW access. (Copyright 1994–1998 Georgia Tech Research Corporation. All rights Reserved. Source: GVU's WWW User Survey *www.gvu.gatech.edu/user_surveys*)

are being used the most. If the scientist insists that they can get by with their VT100 terminal and a text based Lynx browser very soon they will be unable to browse sites that are visually rich or rely on Java scripts or corba interfaces. It is clear from the latest survey results that the most used widely used computing platform is Windows 95 (*Figure 3*). No doubt this is partly due to the popularity of Microsoft Internet Explorer which comes bundled with the operating system. The browser wars between Netscape and Microsoft have already led to legal battles in the American courts.

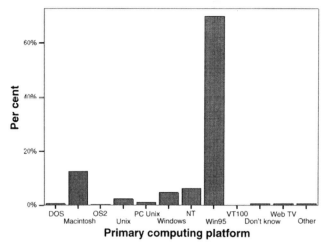

Figure 3 Primary computing platform. (Copyright 1994–1998 Georgia Tech Research Corporation. All rights Reserved. Source: GVU's WWW User Survey *www.gvu.gatech.edu/user_surveys*.)

2.1 Networking in Europe

When it comes to networking not all countries are created equal. The EMBnet organization has developed a service called 'Network Performance monitoring in EMBnet'. This project has monitored the efficiency of networking throughout Europe between the EMBnet nodes. The URL that gives the results from this project is *http://www.cmbi.kun.nl/Ping/*.

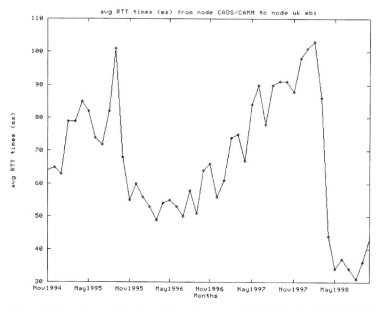

Figure 4 Average round trip times times in ms from EmbNet node CAOS/CAMM in the Netherlands to the EBI in the UK.

In 1995, a *Network Usage and Quality Advisory Group* of the Dutch Network organisation SURFnet, defined '*an upper RTT (Round Trip Time) limit of 125 msec. without packet loss*' as a minimum QOS (Quality of Service) level for interactive on-line work. The RTT values from the Dutch Embnet node to the EBI can be represented by the following graph.

This shows that the RTT from the Netherlands to the EBI in the UK has consistently been below the recommended time of 125 msec, which means scientists from the Netherlands, should have no difficulties in contacting the EBI web server. It is interesting to note that the results from October 1998 show that of the thirty-two nodes monitored, twenty-two have a RTT of less than 125 msec. This must surely be good news for networking within Europe (*Table 1*).

It is essential that research workers learn to use the services provided for them within their own countries. Penalties are always paid when you network across international borders. It would seem that the more borders you cross, the less efficient the network becomes. However, networking within your own country is more efficient because more often than not a basic infrastructure already exists between the major universities. In the mind of the molecular biologist, however, Mecca is either at the NCBI or EBI and that is the direction they religiously point their browsers to, only to suffer frustration when they cannot get their work done due to poor bandwidth and increased traffic directed towards these sites. For this reason EMBnet tries to co-ordinate their activities so that all the EMBnet nodes provide easy access for database query and retrieval. Many of the EMBnet nodes use a mirror package to update their databases on a daily basis, via remote ftp from the databases stored on the EBI anonymous ftp server at *ftp://ftp.ebi.ac.uk/pub/databases*.

The major databases such as EMBL or SWISS-Prot are then indexed at the EMBnet nodes and can be queried with the SRS package. SRS which was developed at EMBL Heidelberg by Thure Etzold, has been adopted by many of the EMBnet nodes throughout Europe and also abroad. SRS is also unique in that it is able to index very many different databases. A list of all EMBNET sites that use SRS is given in *Table 2*.

2.2 The way we were ... e-mail servers for sequence retrieval

Networking for the biologist has a very short history and many of the services developed in the eighties are still in use today. Indeed for people with very bad network connections the use of E-mail servers is still the preferred method of obtaining sequences or running homology similarity searches such as FASTA or Blast. The main depositories for sequence data are are found in the UK at the European Bioinformatics Institute (EBI), and at the National Centre for Biotechnology Information (NCBI) in the United States. In addition these two institutes collaborate with the DDJBB in Japan.

Both the EBI and NBCI run e-mail servers that will allow you to retrieve sequences via e-mail. To obtain information on how to run the e-mail server at

Table 1 Table of EMBNET Internet PING results from 1–30 September 1998. These give network response times from the CAOS/CAMM centre in Nijmegen (The Netherlands) to all of the European EMBNET sites. The main figures show the average, minimum, and maximum RTT times.

Node	% Loss	Avg RTT (ms)	Min RTT (ms)	Max RTT (ms)
nl_caoscamm	0	1	1	30
uk_ucl	0	32	20	334
de_embl	98	33	26	86
be_ben	3	34	19	567
uk_hgmp	0	35	22	290
uk_ebi	1	36	22	251
uk_sanger	1	37	23	394
uk_seqnet	2	42	30	289
ch_expasy	5	42	28	328
ch_isrec	3	43	28	258
de_dkfz	5	46	31	336
fi_csc	2	50	43	247
no_bio	0	52	45	156
se_bmc	0	56	44	564
fr_infobiogen	1	56	31	249
dk_biobase	5	59	49	654
at_biocenter	11	64	49	279
es_cnb	3	65	44	415
fr_genethon	1	65	35	851
de_mips	7	67	41	1091
ie_incbi	4	74	39	558
hu_abc	6	87	50	542
it_icgeb	4	132	58	2299
it_cnr	8	176	82	1214
us_ncbi	3	181	106	2804
gr_imbb	15	188	76	808
il_inn	95	275	149	521
pt_pen	3	372	103	3724
pl_ibb	3	588	560	876
au_angis	15	669	433	1828
za_sanbi	9	816	687	2084
cn_peking	34	904	742	3160

EBI you simply send a e-mail message to *netserv@ebi.ac.uk* and include in the main body of the message the word help and full instructions will be sent via e-mail on how to operate the service.

A similar method for sequence retrieval is employed by the NCBI and the e-mail query system utilizes the Entrez retrieval system that they have developed

Table 2 A List of the EMBNET nodes and some other sites around the world which support SRS

WEHI, Melbourne, Australia
Vienna Biocenter EMBnet Node, Vienna, Austria
Belgian EMBnet Node (BEN), Brussels, Belgium
DBBM-IOC, Fiocruz, Rio de Janeiro, Brazil
CBR-NRC, Halifax, Canada The Genome Mine, Base4 Bioinformatics, Canada
CBI EMBnet Node, University of Beijing, China
CSC, Otaniemi, Espoo, Finland
INFOBIOGEN, Villejuif, France Institut Pasteur, Paris, France
LBMRPM INRA/CNRS, Auzeville, Toulouse, France
DKFZ, Heidelberg, Germany
EMBL, Heidelberg, Germany
GBF, Braunschweig, Germany
MIPS–MPG/GSF, Martinsried/Munich, Germany
Bioinformatics Centre, University of Pune, India
INCBI EMBnet Node, Dublin, Ireland
Weizmann Institute BCD, Rehovot, Israel
CNR EMBnet Node, Bari, Italy
CRISCEB, Second University of Naples, Italy
IVR, Kyoto University, Japan
Biotek EMBnet Node, Oslo, Norway
IBB-PAS EMBnet Node, Warsaw, Poland
IGC EMBnet Node, Oeiras, Portugal
SRCG, Novosibirsk, Siberia, Russia
BIC, National University Hospital, Singapore
CNB EMBnet Node, Madrid, Spain
Biomedical Centre (BMC), Uppsala, Sweden
ExPASy, Geneva, Switzerland
CAOS/CAMM Center, Nijmegen, The Netherlands
RIGEB-MRC, Gebze, Kocaeli, Turkey
Adlib, CAB International, Wallingford, UK
EMBL-EBI, Hinxton, Cambridge, UK
HGMP-RC, Hinxton, Cambridge, UK
MBDC Oxford, Oxford University, UK
SEQNET EMBnet Node, Daresbury, UK
Sanger Centre, Hinxton, Cambridge, UK
IUBio, Indiana University, USA

for their website. Many people would argue that getting sequence via e-mail is old-fashioned technology. It is primitive in that it only delivers simple ascii-formatted text. However the e-mail query server at the NCBI is clever enough to be able to return the sequence to you in a variety of different formats including GenBank, FASTA, or Html.

Protocol 1

Using *netserv@ebi.ac.uk* for the retrieval of sequences and software

To request:

- Specific help on the sequence databases such as EMBL or SWISS-PROT
- General help on software
- The sequence with accession number X03392 (nucleotide)

- The sequence called PIP03XX (nucleotide)
- The sequence called WAP_MOUSE (protein)
- The sequence submission form

You would write the following commands directly into the body of an e-mail message:

HELP NUC
HELP PROT
HELP SOFTWARE
GET NUC:PIP03XX
GET NUC:X03392
GET PROT:WAP_MOUSE
GET DOC:DATASUB.TXT

and then mail the commands to the e-mail address *netserv@ebi.ac.uk* You would then receive the results back in your mailbox via e-mail.

It is often more convenient to shoot off a query by e-mail and get an answer within a few minutes than it is to struggle with trying to access a website that has bandwidth problems. The address for the NCBI e-mail server is at *query@ncbi.nlm.nih.gov*. To receive full instructions on how the server works just send an e-mail message to *query@ncbi.nlm.nih.gov* and in the main body of the message type the word help. I have often found that people who have used an e-mail server generally have a better understanding of databases and sequence retrieval than those who have only used a WWW interface.

Protocol 2

Using *query@ncbi.nlm.nih.gov* for sequence retrieval

Examples:

DB n
UID U30150

Will search the nucleotide database for an entry whose accession number is U30150. Since no DOPT line is present, the record will be displayed the record in the default GenBank format.

DB n
UID U30150,U30153
DOPT f

Will search the nucleotide database for entries whose accession numbers are U30150 and U30153, and display them in FASTA format.

DB m
UID 88055872
DOPT r
HTML
Will search the MEDLINE database for the record with MEDLINE UID 88055872 and display it in MEDLINE Report format. Send the results in HTML format for viewing through a WWW browser.

DB p
UID sp|P11598|
DOPT m
Will search the protein database, using a FASTA formatted UID, to retrieve the entry whose Swiss-Prot accession number is P11598, and display the MEDLINE links for that protein record as document summaries.

2.3 Similarity searches via e-mail

The two most popular e-mail servers dealing with similarity searches are Blast from the NCBI. and FASTA from EBI. For help regarding these e-mail servers you can send an e-mail message to either *blast@ncbi.nlm.nih.gov* or *fasta@ebi.ac.uk* and complete instructions on how to formulate an e-mail message to be processed by these servers will be returned to you via e-mail. Again it should be stressed that once you understand how to compose an e-mail message to submit a Blast query via E-mail, then you can be more discriminating when you are asked to repeat the procedure via the WWW. As it is most people just opt for the default parameters and never experiment with different options.

Protocol 3

Blast similarity search e-mail server at NCBI

To submit a Blast similarity search at the NCBI a e-mail message should be composed as follows.

From: rab.c.nesbit@goven.com Tue Jul 28 21:36:38 1998
Date: 28 Jul 1998 21:29:02-EDT
To: blast@ncbi.nlm.nih.gov
Subject:
PROGRAM blastn
DATALIB month
EXPECT 0.75

BEGIN
>XYZ012 mygene XYZ
tgcttggctgaggagccataggacgagagcttcctggtgaagtgtgtttcttgaaatcat

The actual search request begins with the mandatory parameter 'PROGRAM' in the first column followed by the value 'blastn' (the name of the program) for searching nucleic acids. The next line contains the mandatory search parameter 'DATALIB' with the value 'month' for the newest nucleic acid sequences. The third line contains an optional EXPECT parameter and the value desired for it. The fourth line contains the mandatory 'BEGIN' directive, followed by the query sequence in FASTA/Pearson format. Each line of information must be less than 80 characters in length. Once the e-mail message has been sent it will be processed automatically at the NCBI and the results returned to your e-mail address once they have been computed.

The BLAST algorithm was developed by the National Center for Biotechnology Information at the National Library of Medicine. The BLAST family of programs employs this algorithm to compare an amino acid query sequence against a protein sequence database or a nucleotide query sequence against a nucleotide sequence database, as well as other combinations of protein and nucleic acid. If you use BLAST as a tool in your published research, the following reference should be cited:

Altschul, S. F., Madden, T. L., Schaffer, A. A., Zhang, J., Zhang, Z., Miller, W., and Lipman, D. J. (1997).Gapped BLAST and PSI-BLAST: a new generation of protein database search programs. *Nucleic Acids Res.*, Sept. 1, **25**(17), 3389.

It used to be that the NCBI exclusively provided access to BLAST but in recent years you can now run BLAST searches from many different sites around the world, which is a clear indication that this programme has become a very popular method for doing homology searches. The fact that it appears in so many places may be due to the fact that it is available for free from the NCBI anonymous ftp server at *ftp://ftp.ncbi.nlm.nih.gov/blast/*

Historically the EBI has always provided homology searches through FASTA. The following reference should be cited when you have used FASTA:

Pearson, W. R. and Lipman, D. J. (1988). Improved tools for biological sequence analysis. *Proc. Natl. Acad. Sci. USA*, **85**, 2444.

Web-based FASTA applications can be found at the EBI *http://www.ebi.ac.uk/fasta3* and at DDJB *http://www.ddbj.nig.ac.jp/E-mail/homology.html*

Protocol 4

FASTA similarity search e-mail server at EBI

EXAMPLE OF A SIMPLE SUBMISSION
PATH mary.doll@goven.com
TITLE My Sequence
LIB swall
SEQ
MMFSGFNADYEASSSRCSSASPAGDSLSYYHSPADSFSSMGSPVNAQDFC
TDLAVSSANFIPTVTAISTSPDLQWLVQPALVSSVAPSQTRAPHPFGVPA
END

The PATH is the e-mail address of the person to whom the results should be sent. TITLE is anything that you want to appear on the subject line of the returned mail. LIB is the database that you want to search against. The sequence itself should be enclosed between SEQ and END.

2.4 Speed solutions for similarity searches

In recent years there has been an increase in the use of specialized hardware for doing similarity searches. Four companies in particular have pioneered this approach, and the turn around time for running a search against the whole of Swiss-Prot has been reduced to around 10 seconds using the Smith–Waterman algorithm.

2.4.1 Time Logic

Time Logic (*http://www.timelogic.com*) from the USA has introduced DeCypher Bioinformatics Accelerators and they have implemented a number of algorithms namely, Gapped BLAST 2 (includes entire heuristic search suite: blastn, blastp, blastx, tblastn, tblastx) PSI-BLAST, Affine Smith–Waterman, FrameSearch, ProfileSearch, ProfileScan, and ClustalW with graphical rendition of dendrogram (Java applet). The WWW query interface for a Smith–Waterman similarity search is shown in *Figure 5* and some results from that search are given in *Figure 6*. Timelogic also have a Blast search running on their hardware at the NCGR at the URL:

http://seqsim.ncgr.org/newBlast.html

2.4.2 Compugen

Compugen have succeeded in introducing the Biocellorator to many pharmaceutical companies to aid them in their search for new and novel drugs. The EBI has a biocellorator, which is online and is available for public use. At the EBI there are two different interfaces to this service. The one provided by Compugen called GeneWeb and a simple custom interface developed at the EBI at *http://www.ebi.ac.uk/bic_sw*. The interface to the BIC-SW at the EBI is very compact

Figure 5 DeCypher interface for Smith–Waterman similarity searches.

and easy to use. Most people just accept the default settings, paste in their query sequences and run the program. The interface page is shown in *Figure 7* and some results are shown in *Figure 8*.

If you are from the Mediterranean area then perhaps it would be more convenient to try the GeneWeb interface from the Weizmann Institute in Israel, which is also open to the public for unregistered users. The URL is (*http://sgbcd.weizmann.ac.il:80/cgi-bin/genweb/main.cgi*)

2.4.3 Paracel FDF

At the Swiss EMBnet node you can find the Paracel Fast Data finder (FDF), which is designed to help bioinformatics departments dramatically increase the rate at which they can find high-scoring potential genomic targets. Paracel claim that GeneMatcher is the first commercially available genetic data analysis system to use custom ASIC technology that can analyse similarities or differences in DNA or protein sequences up to 1000 times faster than traditional computer systems.

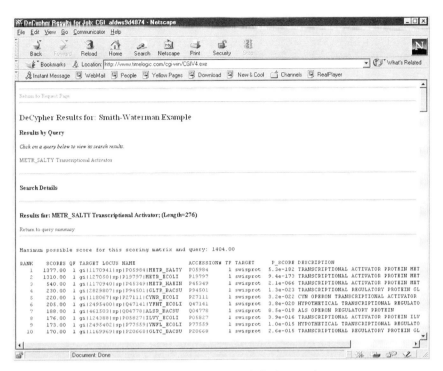

Figure 6 Results from Decypher Smith–Waterman similarity search.

Competition to discover novel genes is of great interest to pharmaceutical companies because if it is possible to identify just one critical target gene then this can result in an application for a patent on a product. Therefore any method that combines speed with sensitivity is a very valuable tool in the hands of the research worker. The main search interface and some results are shown in *Figures 9* and *10*.

3 Sequence retrieval via the WWW

If you are in a country with a poor Internet connection then working with E-mail servers for the retrieval of sequences is often the best option. However, there are many excellent servers in different parts of the world and they should not be ignored, even if you do live on the other side of the planet. Your geographical location should be the first consideration when accessing a remote site. It is best to access a site that is in close proximity. Two of the most popular services for sequence retrieval are Entrez from the NCBI and SRS from the EBI. However there are other options available and if you are in the Pacific rim area then it might be worthwhile to look at the services offered by DDBJ in Japan *http://www.ddbj.nig.ac.jp/searches-e.html* or the Maestro service from the National Centre for Genomic Research (NCGR) *http://www.ncgr.org/gsdb/maestro/index.html* on the West Coast of the USA.

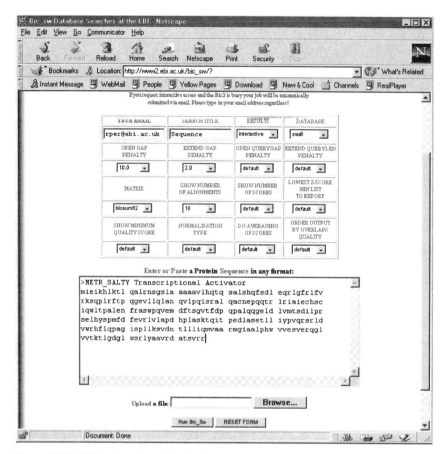

Figure 7 BIC-SW interface for performing Smith–Waterman homology searches.

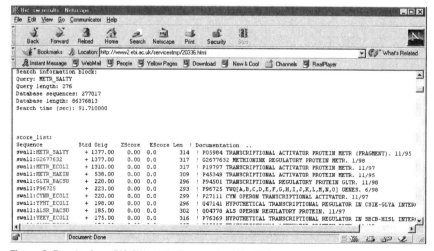

Figure 8 Results from BIC-SW similarity search at the EBI.

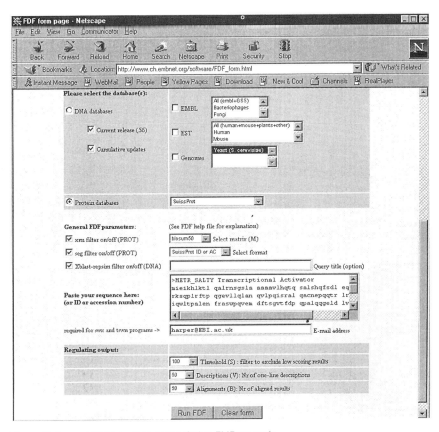

Figure 9 Paracel FDF interface at the Swiss EMBnet node.

3.1 Entrez from the NCBI

The NCBI is the only place in the world where you will find the Entrez service and it concentrates on a few databases namely, nucleotide sequences, amino acid sequences, 3-D structures, Genomes, Taxonomy, and Literature-PubMed *http://www.ncbi.nlm.nih.gov/Entrez/*. One of its strengths is that it provides access to PubMed and this is a key factor in its popularity and success. Effective August 3, 1998, NLM implemented a system enhancement that dramatically increases the speed of the system. This redesign of the way PubMed stored and retrieved information will improve users search time—a search that previously took approximately 18 seconds to run in PubMed now runs under 2 seconds.

In Entrez you select the database you wish to query, for example the protein database and then you are allowed to string a number of keywords together, like 'Rhizobium Ausubel nodulation' and those entries that meet the criteria will be displayed. An example is shown in *Figure 11*.

3.2 SRS from the EBI

SRS *http://srs.ebi.ac.uk/* is a very powerful tool for querying databases and it would seem to be the preferred querying system within Europe. You select a database

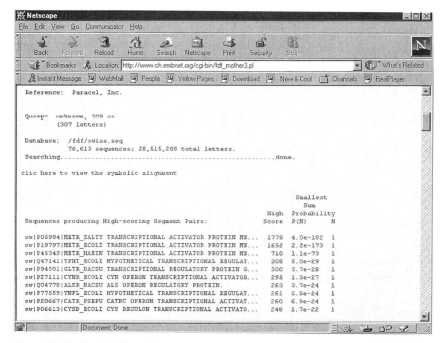

Figure 10 The first few lines of some search results from an FDF search at the Swiss EMBnet node.

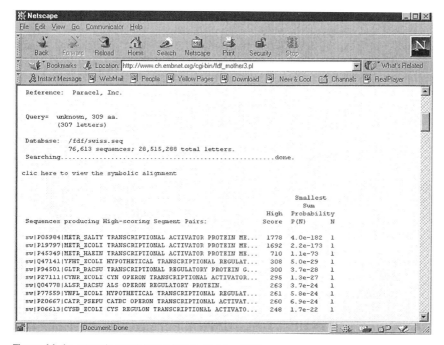

Figure 11 An example query using Entrez and the NCBI.

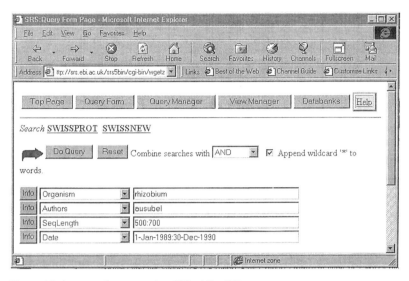

Figure 12 An example query using SRS at the EBI.

and fill in your search criteria as keywords. For example in *Figure 12*, we see a sample query using the fields Organism (Rhizobium), authors (Ausubel), Seq-Length (a range 500:700) and the date (a range 1-Jan-1998:30-Dec-1990).

SRS will then display two hits in Swiss-Prot for that particular year with a sequence range between 500 and 700 (*Figure 13*). It should also be noted that SRS also gives the possibility to launch an application such as BLAST or FASTA for any of the sequences that you care to select. You may also select different views of a sequence. For example the FASTA format, which then allows you to launch a multiple sequence alignment using ClustalW, directly from within SRS. This

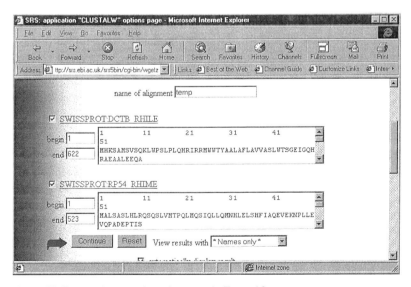

Figure 13 The results page from the query in Figure 12.

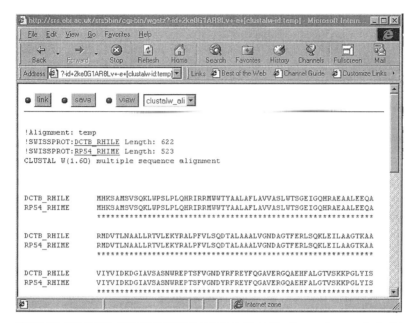

Figure 14 The results from launching clustalw as an application on the results in Figure 13.

method is a great time saver since there is no need to cut and paste your sequence into a separate ClustalW application (*Figure 14*).

4 Submitting sequences

Not only does the research worker want to query, retrieve and analyse sequences, occasionally they also want to submit their own sequences to the databanks be it GenBank or EMBL. The three major organizations that collect sequence information work in collaboration with each other so that sequences entered into GenBank are transferred daily by FTP to both EBI and DDBJ (and vice versa) in an attempt to keep the major databases synchronized.

At any given time the three institutes are continually swapping data so it is a false idea to believe that any one database is more current than the other. All three institutes have online methods of submitting sequence data through the Web. The NCBI were the first to come online with BANKIT. The EBI then followed with WEBIN and the Japanese at DDBJ have Sakuara.

It should also be noted that the NCBI developed a stand-alone programme for MAC's, PC's, and Unix called SEQUIN that allows the end-user to enter their data from a personal computer and to send the submission via e-mail or to simply post the disk to the appropriate institute where it is then uploaded into the database. Sequin is strongly recommended if you have bulk submissions to make.

4.1 Bankit at NCBI

Bankit is convenient for quick submission of sequence data to the NCBI. BankIt allows you to enter sequence information into a form, edit as necessary, and add biological annotation (e.g. coding regions, mRNA features). BankIt transforms your data into GenBank format for you to review and when your record is completed, it can be submitted directly to GenBank. You have the option of adding information by using text boxes to describe in your own words the source of the sequence and its biological features. The entry screen from Bankit is shown in *Figure 15*. The GenBank annotation staff reviews the submitted textual information, incorporates it into the appropriate structured fields, and returns the record by e-mail for your review.

4.2 Sequin from NCBI

Sequin is stand-alone program for the MAC, PC/Windows and UNIX. Sequin is an interactive, graphically oriented program based on screen forms and controlled vocabularies that guide you through the process of entering your sequence and providing biological and bibliographic annotation. Sequin is designed to simplify the sequence submission process and to provide graphical viewing and editing options. This program is optimal for submitting multiple sequences, mutation studies, phylogenetic sets, population sets, and segmented sets. It incorporates

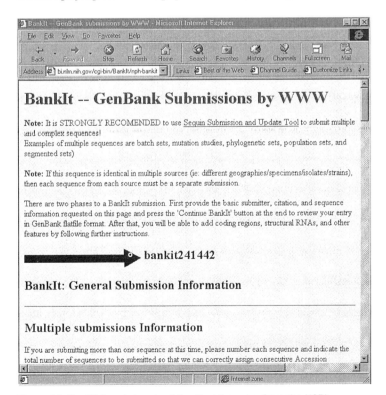

Figure 15 The Welcoming page of the Bankit service from the NCBI.

Figure 16 The welcoming screen of the Sequin data submission program from the NCBI.

robust error checking and accommodates very long sequences and complex annotations.

Although Sequin has been implemented by the NCBI, the opening screen allows you to select which database you would like to submit you sequence to be it GenBank, EMBL, or DDBJ. Usually when a sequence is submitted there may be a process whereby the submitter has to be in contact with the annotators of the sequence by telephone to clarify certain details. Therefore it is wise to choose a submission centre in your geographical region if you want to avoid long distance telephone calls. A screen capture of Sequin is shown in *Figure 16*.

Once you have completed the submission depending on which database you have selected at the beginning you will be prompted to send an e-mail to *gbsub@ncbi.nlm.nih.gov* for NCBI, *datasubs@ebi.ac.uk* for EMBL or *ddbjsub@ddbj.nig.ac.jp* for DDBJ. Sequin runs on Macintosh, PC/Windows, and UNIX computers. The program itself, along with its on-line help documentation, is available by anonymous FTP from the *EBI (UK)* at ftp://ftp.ebi.ac.uk/pub/software/sequin/ or from the *NCBI (USA)* at ftp://ncbi.nlm.nih.gov/sequin/ A useful FAQ to help you if you run into problems during submission can be found at. http://www.ebi.ac.uk/~sterk/sqndocs/index.html

4.3 Webin from EBI

The EBI WWW tool (WebIn) guides the user through a sequence of WWW forms allowing the user to submit sequence data and descriptive information in an interactive and easy way (see *Figure 17*). All the information required to create a database entry will be collected during this process:

(a) Submitter information.

(b) Release date information.

(c) Sequence data, description, and source information.

(d) Reference citation information.

(e) Feature information (e.g. coding regions, regulatory signals etc.).

Data submissions are usually processed within two working days of receipt and the authors are sent notification of their accession number(s). Authors will be asked whether their submitted data can be made available to the public immediately or whether they should be withheld until an author-specified date. Data are never withheld after publication.

Once a database entry has been created from a submission, a copy is sent to the submitter for their reference and for comments or corrections. However, it often happens that the entry is correct when it is created but, with the passage of time, becomes out of date. The authors may make corrections to the sequence itself, or may discover new features of the sequence. Since such findings are often not published, the only way to keep entries correct and up to date is if the authors communicate their new findings to the database. At the EBI this can be done by completing an update form available from the Anonymous FTP, site FTP.EBI.AC.UK in the file: pub/databases/embl/release/update.doc or via the WWW at the URL http://www.ebi.ac.uk/ebi_docs/update.html.

A new service that has been instituted at EBI is scanning for vectors before

Figure 17 A page from the Webin service of the EBI.

211

submitting your sequence. You are able to check your sequence data prior to submission for potential vector contamination by running a BLASTN search against EMVEC, a vector database containing information on more than 2000 vectors from the EMBL/GenBank/DDBJ Database SYN(thetic) division. The results will list sequences producing significant alignments and associated information like vector name, score, alignment, etc. The EBI suggests that you remove vector contamination from your sequence data before submitting to the database.

4.4 Sakura from DDJB

SAKURA is a web-based DNA data submission system for DDBJ. The URL for SAKURA is http://sakura.ddbj.nig.ac.jp which can be accessed from the DDBJ Home Page (http://www.ddbj.nig.ac.jp). You can select either the English or Japanese version. However, data input must be done in English only, regardless of language version selected . SAKURA allows you to save your document before completion and submit multiple sequences sequentially (see *Figure 18*).

5 Conclusions

Historically there has been a collaboration between EBI, NCBI, and DDBJ. These three sites are still the only places that have the infrastructure set up to handle

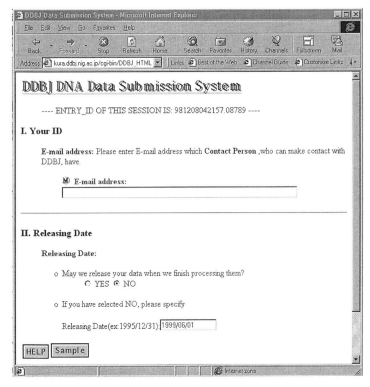

Figure 18 A page from the SAKURA DNA submission system from DDBJ.

the submission of nucleotide sequences to the databases, be they EMBL or Genbank or DDBJ. For this reason they are also looked upon as the only places where you can do queries and retrieval, or perform homology searches, or multiple sequence alignments. This is no longer true and with the advent of EMBnet, many of the national nodes are able to supply services that are not offered by the major centres. These three major centres have a policy of making all of their databases publicly available, and when distributed network of databases exists in many different parts of the globe then it can only be for the benefit of molecular biologists worldwide.

References

General references to articles about biological services on the internet.

1. Aldhous, P. (1993). Managing the genome data deluge. *Science*, **262**, 502.
2. Altschul, S. *et al.* (NCBI) (1994). Issues in searching molecular sequence databases. *Nature Genet.*, **6**(Feb), 119.
3. Appel, R. D., Sanchez, J.-C., Bairoch, A., Golaz, O., Ravier, F., Pasquali, C., *et al.* (1996). The Swiss-2DPAGE database of two-dimensional polyacrylamide gel electrophoresis, its status in 1995. *Nucleic Acids Res.*, **24**(1), 180.
4. Ashburner, M. and Goodman, N. (1997). Informatics—genome and genetic databases. *Curr. Opin. Genet. Dev.*, **7**, 750.
5. Bairoch, A., Bucher, P., and Hofman, K. (1996). The Prosite Database, its status in 1995. *Nucleic Acids Res.*, **24**(1), 189.
6. Bairoch, A. and Apweiler, R. (1996). The Swiss-Prot Protein sequence data bank and its new supplement Trembl. *Nucleic Acids Res.*, **24**(1), 21.
7. Bairoch, A. (1996). The ENZYME Data Bank in 1995. *Nucleic Acids Res.*, **24**(1), 221.
8. Bairoch, A. (1991). SEQANALREF: a sequence analysis bibliographic reference databank. *Comput. Appl. Biosci.*, **7**(2), 268.
9. Bleasby, A., Griffiths, P., Harper, R., Hines, D., Hoover, K., Kristofferson, D., *et al.* (1992).Electronic communications and the new biology. *Nucleic Acids Res.*, **20**(16), 4127.
10. Coulson, A. (1994). High performance searching of biosequence databases. *Trends Biotechnol.*, **12**, 76.
11. Fuchs, R. (1994). Sequence analysis by electronic mail: a tool for accessing Internet e-mail servers. *Comput. Appl. Biosci.*, **10**(4), 413.
12. Gershon, D. (1997). Bioinformatics in the post-genomic age. Careers and recruitment article. *Nature*, **389**, 417.
13. Gershon, D. (1995). The boom in bioinformatics (employment review). *Nature*, **375**, 262.
14. Harper, R. (EBI). (1995). World Wide Web resources for the biologist. *Trends Genet.*, **11**(6), 223.
15. Holm, L. and Sander, C. (1996). The FSSP database: fold Classification based on structure-structure alignment of proteins. *Nucleic Acids Res.*, **24**(1), 206.
16. Marshall, E. (1996). Hot property: biologists who compute. *Science*, **272**, 1730.
17. O'Donnell, C. (1994). Obtaining software via INTERNET. *Methods Mol. Biol.*, **24**, 345.
18. Peitsch, M. C., Wells, T. N., Stampf, D. R., and Sussman, J. L. (1995). The Swiss-3DImage collection and PDB-Browser on the World-Wide Web. *Trends Biochem. Sci.*, **20**(2), 82.
19. Pietrokovski, S., Henikoff, J. G., and Henikoff, S. (1996). The Blocks database a system for protein classification. *Nucleic Acids Res.*, **24**(1), 197.

20. Roberts, R. J. and Macelis, D. (1996). REBASE—restriction enzymes and methylases. *Nucleic Acids Res.*, **24**(1), 223.

21. Rodriguez-Tomé, P., Stoehr, P., Cameron, G. N., and Flores, T. P. (1996). The European Bioinformatics Institute (EBI). *Nucleic Acids Res.*, **24**(1), 6.

22. Smith, T. F. (1990). The history of the genetic sequence databases. *Genomics*, **6**(4), 701.

23. Stoehr, P. J. and Omond, R. A. (1989). The EMBL Network File Server. *Nucleic Acids Res.*, **17**(16), 6763.

24. Williams, G. W. and Gibbs, G. P. (1990). Automatic updating of the EMBL database via EMBNet. *Comput. Appl. Biosci.*, **6** (2), 122.

Chapter 10

SRS—Access to molecular biological databanks and integrated data analysis tools

D. P. Kreil and T. Etzold

EMBL-European Bioinformatics Institute, Wellcome Trust Genome Campus, Cambridge, UK.

1 Introduction

This first section gives an introduction to SRS. Section 2 (A user's primer) is a tutorial that demonstrates basic tasks: simple database queries, exploiting links between databases, exploration of results, and launching analysis tools. Section 3 (Advanced tools and concepts) builds on the skills imparted by the tutorial, and shows how to refine queries, create custom views on data, and use distributed SRS resources. Section 4 (SRS server side) introduces aspects of using a local SRS installation, and outlines how to take advantage of one. Section 5 (Where to turn for help) suggests where to turn to, if this chapter does not address a particular question or problem.

1.1 SRS fills a critical need

Everyone who has faced the problem of finding particular information (e.g. experimental results) in the printed literature knows to value computer-readable storage of data. Besides allowing advanced methods of data retrieval and visualization, computerized storage also facilitates systematic combination of data from multiple sources, comparative studies, and methodical application of analysis tools to large sets of data. The rapid growth of available molecular biological databases now gives us access to an unprecedented amount of information by computer.

The increasing specialization of research and the resulting fragmentation of molecular biology is clearly reflected in the choice of databases, which are many and varied in content type and form. The wide differences that can be observed in the database formats and technologies employed are partly caused by the different nature of the results gathered, and cannot solely be blamed on an indifference on the part of the database designers towards efforts at standardization

While adhering to present standards is certainly to be advocated, it would often significantly impede the progress of individual research groups. Also, many databases, and consequently the respective data formats, have grown over history.

The large number of different databases alone is a serious problem for knowing where to turn for specific information. DBCAT, a manually maintained catalogue of databases, currently lists around 400 databases relevant to biology (1). The variation in database formats and technologies that need to be dealt with yet adds significantly to an already complex task of fruitfully accessing the available data. The technical difficulties encountered particularly obstruct more sophisticated leverage of data, such as systematic application of analysis tools, or bringing data from different areas together. Even though many databases directly or indirectly reference data stored elsewhere, these links are difficult to exploit, owing to the large differences between individual database implementations. While a standardization of formats, or at least an agreed upon interface for database interconnectivity, would greatly alleviate these problems, none of the attempts so far has achieved the hoped-for degree of acceptance (2).

SRS has evolved to overcome many critical technical difficulties in database access, and the integration of databases, and analysis tools. Already in earlier versions SRS was considered a promising approach to the problems it set out to solve—respected as 'the paragon of connectivity' (3). The current version SRS-5.1 now offers an automatically maintained catalogue-database of over 350 different databases and their documentation to help users identify databases of interest and locate an appropriate server (4). Users can access all these databases through the uniform SRS query interface on the Web. The interface is simple to use, yet allows complex queries, including the use of logical operators, combining fields from multiple databases, and following implicit and explicit links between databases. Results can be browsed in various views, combining data from several databases. Database entries can be passed to analysis tools such as sequence similarity search programs, restriction map analysis, or phylogenetic algorithms. The results generated can be used in further queries or analyses.

1.2 History, philosophy, and future of SRS

The SRS system started out as a Sequence Retrieval System (5) that employed sophisticated parsing and indexing of database text files (a parser is a computer program that breaks down text into recognized strings of characters for further analysis). Plain text is the lingua franca of data exchange. Whether a researcher has entered experimental results into a spreadsheet, or a database centre maintains a repository of complex data structures using an advanced database management system, everyone can provide a dump of their data in the form of plain text files. SRS does not convert these files, but rather leaves the original data in place and reads items directly from there, preserving the original context when the complete entry is viewed. The parsed items are used for display of selected details of a database record, and for the construction of index files that allow efficient data retrieval (6) and searchable links (7) between database entries. SRS was first applied to databases with information on protein or nucleotide sequences. It has since developed into a much more general data-retrieval tool. It is used, for example, for bibliographical databases, hierarchically structured databases like taxonomies, or clinical data on mutations.

Besides giving researchers the freedom to store information the way they want to, an approach centred on the use of plain text files benefits from a medium that is portable across different computer architectures and is usually easy to read for humans. In the integration of distributed resources, this can be quite helpful in resolving conflicts that may be triggered by the individually evolving components.

SRS parsers, definitions of database structures and interfaces to analysis tools, and a considerable amount of the SRS core functionality itself are written in *Icarus*, a language especially designed for that task. Icarus is used to define descriptive data structures (*meta*-data) which are extensively used to represent SRS concepts like database structures. Icarus code in parsers is interpreted. These two design features allow rapid development: using meta-data reduces the amount of code to be written, and interpreted parsers can be modified and tested fast—it often takes just a few hours to integrate a new database starting from scratch. Not only has this approach proven to scale very well with the number of databases to integrate, the flexible combination of a recursive breakdown of structure and the powerful in-place processing of parsed data that is characteristic of SRS parsers handles even very complex database formats elegantly. Developers who have used an interpreted language before will have little difficulty adapting to Icarus. We think this is reflected by its ready acceptance by SRS server maintainers around the world who have added well over hundred databases to the system.

Icarus is evolving towards a general-purpose object-oriented programming language with special support for recursive lazy parsing and definition of meta-data structures. We expect that it will play a central role in scripting and large-scale data analysis in the future of SRS.

SRS integrates and interfaces to databases and analysis tools of other researchers, and it is also often used as an engine for database access in other systems like OPM (8), and BioKleisli (9)—'wrap and be wrapped'! Communication with an SRS system can be through the World Wide Web (the most popular form of access), from the Unix command line, from within an Icarus script, or using the C application programming interface (API). Current new developments include prototypes of a Corba server, and of a Perl API. Adding support for other language API's (e.g. for Java via JNI) would be straightforward.

2 A user's primer

This section will introduce core SRS functionality with simple step-by-step instructions. It concludes with an overview of the screens covered in the primer. We suggest that new users follow the examples given below. The concepts are much more readily understood with some hands-on experience.

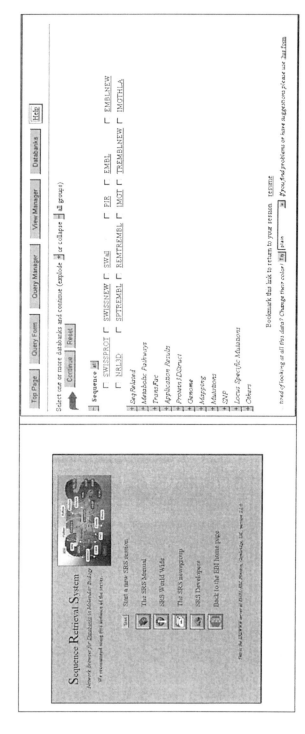

Figure 1 (a) The SRS Home Page at EMBL-EBI, http://srs.ebi.ac.uk/. (b) The 'Top Page': Select the databases to query.

2.1 A simple query

SRS is a very powerful and complex system. Nevertheless, the basic features are easy to use, and its interface also caters for the not so experienced user. *Protocol 1* is a guide to performing a simple query and should be accessible for absolute beginners.

Protocol 1

Performing a simple SRS query

1   Connect to an SRS server, e.g. *http://srs.ebi.ac.uk/* at EMBL-EBI (Figure 1a).

2   Press the 'Start' button to begin a new SRS session.[a]

3   Select one or several[b] databases to query, e.g. SWISSPROT. Database names are displayed in groups. If necessary, expand and collapse groups by clicking on the '+' and '-' buttons respectively (*Figure 1b*).

4   Press the 'Continue' button to get the Query Form (*Figure 2*).

5   Select the field to query in the drop-down box, e.g. 'Description'.

6   Enter the term to search for, e.g. 'Tetracycline' (without the quotes). By default, the wildcard '*' will be appended, thus also matching 'Tetracyclines' etc. Deselect this option for a verbatim search.[c]

7   Press the 'Do Query' button to submit your query.

[a] SRS will store your work. If you plan to come back to it, add the 'resume' link from this page to your bookmarks. At EMBL-EBI, an SRS work session is deleted when it has not been used for two consecutive days. Other public sites may have similar mechanisms in place.

[b] When several databases are selected, only fields present in all the selected databases are available in the Query Form.

[c] Avoid using wildcards when not needed since this considerably speeds up the search.

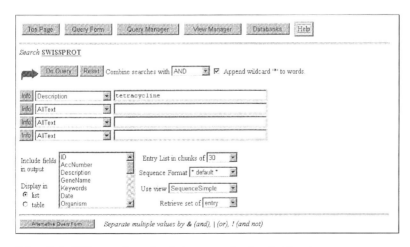

Figure 2 An SRS query form showing an example of a simple query.

If the list of results is long, the user can page through the list using the links at the bottom of the screen ('go to entries in chunk [. . .]'). Every item in the list of results has an identifier composed of database name and entry-i.d., such as 'SWISSPROT:LPTR_BACST'. They are hyper-linked to the respective complete entries.

To submit another query of the same database(s), press the 'Query Form' button and continue with *Protocol 1*, step 5. For a new query, press the 'Top Page' button and go to Protocol 1, step 3. See *Figure 6* for more options.

2.2 Exploiting links between databases

In contrast to many other systems, SRS allows queries that search so-called *links*. The meaning of links between entries depends on the databases involved, yet it is usually evident. For example, an entry *E* in SWISS-PROT is linked to those EMBL entries which contain nucleotide sequences that belong to the protein described by the entry *E*. Similarly, a SWISS-PROT entry *E* is linked to those PDB entries which hold a 3D-structure for the protein described by the entry *E*. EMBL entries are linked to updates in EMBLNEW that supersede the original entries in EMBL. The example given in the next protocol assumes that the reader has followed *Protocol 1*.

Protocol 2

Applying a link query to selected entries

Selecting a set of entries

1 Start from a page showing the results of a query, e.g. after following *Protocol 1*. Performing a simple SRS query.

2 Mark the tick boxes next to the entries you wish to include or exclude. Use the radio button to choose to operate on the 'selected' or on 'all but selected' entries. For example, chose 'selected' and mark entry 'SWISSPROT:TCRB_BACSU'.

Performing a link query

1 Press the 'Link' button.

2 Select the database(s) to link to [cf. *Protocol 1*, step 3], e.g. EMBL.

3 Press the 'Continue' button to ask for the list of results.

Protocol 2 shows how to apply a link query to a selected set of entries. The example query retrieves an EMBL entry that contains the nucleotide sequence of the gene that encodes the protein described by 'SWISSPROT:TCRB_BACSU'. In addition, entries that hold larger stretches of genomic DNA containing this gene are retrieved (such as the complete genome of *Bacillus subtilis*).

2.3 Using Views to explore query results

There are several pre-defined Views, which present various aspects of a list of results. Besides the Views that display just the entry-id's, or the complete entries, the selection of Views available is database dependent. Protocol 3 shows how to display a set of selected entries in a specific View. The example given in the next protocol assumes that the reader has followed *Protocol 1*.

Protocol 3

Displaying selected entries with one of the pre-defined views

Selecting a set of entries

1 Start from a page showing the results of a query, e.g. after following *Protocol 1*.

2 Mark the tick boxes next to the entries you wish to include or exclude. Use the radio button to choose to operate on the 'selected' or on 'all but selected' entries. For example, chose 'selected' and mark the entries 'SWISSPROT:LPTR_BACST', 'SWISSPROT:LPTR_BACSU', 'SWISSPROT:TCR1_ECOLI', and 'SWISSPROT: TCR3_ECOLI'.

Changing the View

1 Select a View from the drop-down menu next to the 'view' button, e.g. 'Sequence Simple'.

2 Optionally change the number of entries to be shown per screen (. . . *entries in chunks of . . .*').

3 Press the 'view' button.

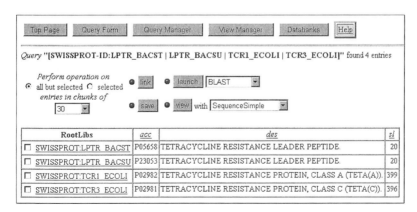

Figure 3 Showing a selection of entries in the View 'SequenceSimple'.

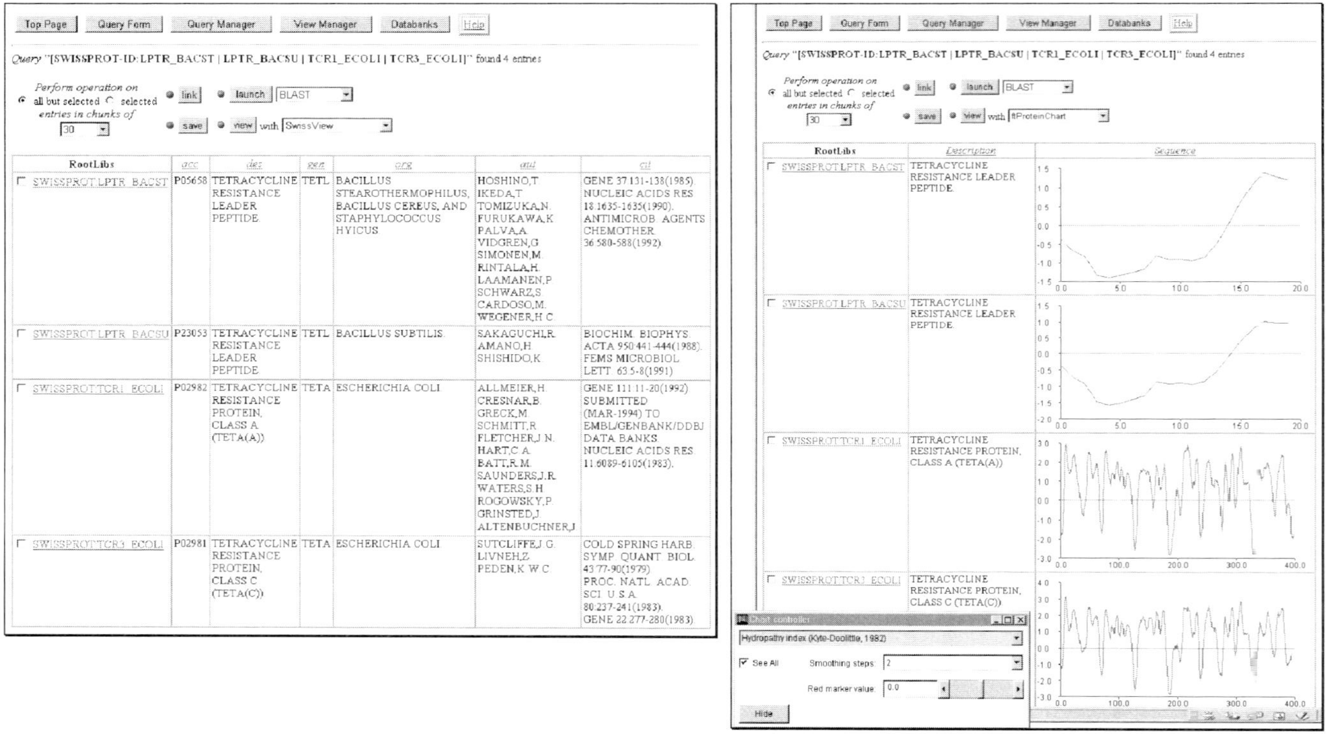

Figure 4 More examples of Views showing the same selection of entries. The screen on the right (b) employs a Java Applet to display various local protein properties as controlled by the user.

Some pre-defined Views present connected fields from multiple databases by linking the displayed entries to other databases and integrating the results. See below for an example.

2.4 Launching analysis tools

Besides browsing results of database field and link queries, external application programs can be used to analyse data. Typical of such applications are database searches by sequence similarity, e.g. BLAST (10), construction of multiple sequence alignments, e.g. CLUSTALW (11), restriction map analysis, and tools predicting various sequence properties (trans-membrane regions, protein secondary structure, etc).

Protocol 4

Launching an external application program for selected entries

Selecting a set of entries

1 Start from a page showing the results of a query, e.g. after a query for 'uroplakin' in the 'Description' field of SWISS-PROT (cf. *Protocol 1*).

2 Mark the tick boxes next to the entries you wish to include or exclude. Use the radio button to choose to operate on the 'selected' or on 'all but selected' entries. For example, chose 'selected' and mark the entry 'SWISSPROT:UPKB_MUSVI'.

Request an application launch

1 Select a tool from the drop-down menu next to the 'launch' button, e.g. 'SW', the Smith & Waterman hardware accelerated search at EMBL-EBI.

2 Press the 'launch' button to get a page of application options.

Adjusting launch parameters and starting the analysis

1 Check and/or amend your original selection of input to the application (for example, by selecting only a fragment of the sequence to be searched for).

2 Adjust the application parameters according to your data and your preferences. Hyper-links offer additional help that is displayed in a separate window to aid data entry (see *Figure 5a*).

3 Consider turning off the 'automatically display result' feature because some browsers do not correctly support it.

4 Press the 'Continue' button.

5 If the 'automatically display result' feature was disabled, wait until the next page has completely loaded; this may take a while. Then follow the 'Display Results' link at the top of the screen.

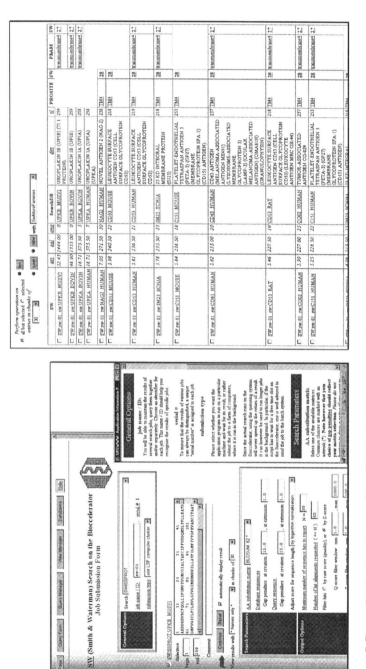

Figure 5 The application launch page of the 'SW' (Smith–Waterman search) application program. Here users can set application parameters. Clicking on a hyper-linked parameter prompts will open a separate window with related help text. This window is re-used to reduce screen clutter. (b) Results of an 'SW' application launch as seen with the view 'SwMoreFamilies'.

Follow *Protocol 4* to launch an external application for selected entries. The results of the application launch as shown in *Figure5b* are an example of a more sophisticated pre-defined View. (See *Protocol 3* on how to choose a pre-defined View.) For each sequence hit by the search, a normalized score (*Z*-score), the raw score, and the number of gaps in the alignment of the query sequence with the hit sequence are displayed. Then each hit is linked to the database searched (SWISS-PROT, in this case) to display the full-length description and the sequence length from that record. Further links to PROSITE and PFAM give information on membership to well classified families. In our example, the search returned related Uroplakin 1A/B sequences as top hits. Examination of the match data together with the family classifications strongly suggests that the longer ranking sequences are not distantly related to the query, but rather scored because they all have a trans-membrane-4 domain. Also, note how the data in PROSITE and PFAM augment each other.

2.5 Overview

Here we give a flow-chart like overview of the interface screens covered by the protocols of this section (*Figure 6*). Users may abort operations and return to the Top Page or a new Query Form by pressing the respective buttons shown at the top of most screens.

3 Advanced tools and concepts

In this section we introduce advanced methods of working with SRS. Users are advised to become familiar with these as soon as they have mastered the basics, as the complexity of the molecular biological data and their storage demands fairly sophisticated data handling skills.

3.1 Refining queries

Due to the nature of molecular biological databases, even simple questions often require a complex query. Here we explain several techniques that help the user go beyond the first simple queries.

3.1.1 How the system works

As in many cases, it helps to know a little about what is happening under the hood. In a nutshell, SRS parsers read text from a file into *entries*. These are then parsed into *fields*, which usually are further decomposed or processed. In particular, it is the parser's job to extract terms to be written into *index* files. These index files make fast queries of database fields possible. Obviously, they also determine what fields are available for queries, and what terms a query can be matched against.

3.1.2 The index browser

Consider searching for human SWISS-PROT entries. SWISS-PROT has an 'organism' field which can be queried for 'human'—or should it be '*homo sapiens*', or

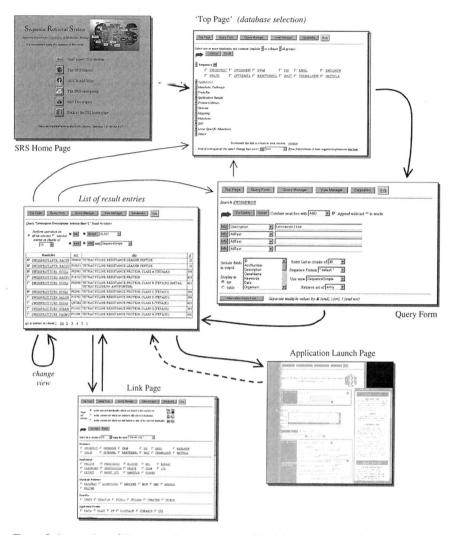

Figure 6 An overview of the pages discussed so far. The thick *arrows* show the path of usual progression as described in the protocols. There may be an intermediate page before the list of results after an application launch.

'*homo sapiens sapiens*', or a combination of all these terms? The index browser is the easiest way to inspect the terms a query can be matched against. Always consult the index browser when working with an unfamiliar database or database field, and when in doubt. *Protocol 5* shows how to do that.

Looking at the returned values, one sees the need to restrict the search to 'human' without a trailing wildcard '*' (to exclude, e.g. the human viruses). Checking the index for terms matching 'homo*', one respectively finds the same number of entries for '*homo*' and '*homo sapiens*' as for 'human'. Queries using logical operators (see below) can be used to check the suspicion that the three terms are equivalent and that it suffices to use only one of them.

Protocol 5

Browsing the index for a database field

1 In the query form, select a database field from the drop-down box, e.g. 'Organism'.

2 Press the 'Info' button next to the selected field name to get the respective Index Browser.

3 Modify the query mask, e.g. enter 'human*' (without the quotes).

4 Press the 'List values' button.

Typical use of the index browser includes:

(a) Checking whether single words or phrases are available from the index for searching. Most free-text fields (such as 'Description') are usually indexed using single words, while fields with a more standardized vocabulary often store longer phrases in the index (e.g. the 'Keywords' field of SWISS-PROT). When a field is indexed using single words, logical operators must be used to combine words. Instead of asking for 'tetracycline repressor', the query must request 'tetracycline' *and* 'repressor' (see below).

(b) Learning about the terms used in fields that have a controlled vocabulary (e.g. the SWISS-PROT Feature Key, 'FtKey').

(c) Learning about special standardized formats used to represent data in indices of particular fields (see the 'Citation' of SWISS-PROT, as an example).

(d) Looking for alternative spellings or words with typing errors using wildcards or a regular expression. SRS regular expressions are delimited by forward slashes, e.g. '/p[0–9] | /', matching 'p' followed by one or more digits.

The index browser also shows documentation for the database field, and status information on the respective index (*Figure 7*).

3.1.3 Complex queries with logical operators

Logical operators are a powerful extension to queries, whether one needs to refine or broaden a search, or it is necessary to combine words of a phrase for searching a field with a single word index. SRS supports the following logical operators:

• '*a* AND *b*' , which requires *both* the terms *a* and *b* to be present—*in queries*: '&';

• '*a* OR *b*', which requires at least one of the terms *a* and *b* to be present—*in queries*: '|';

• '*a* BUTNOT *b*', which requires the term *a* to be present and *b* to be absent—*in queries*: '!'.

The short forms of the logical operators ('&' '|', and '!') can be used when typing in the query form. Please note that parentheses cannot be used for grouping,

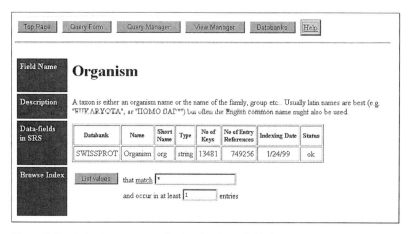

Figure 7 The index browser page for the database field 'Organism'.

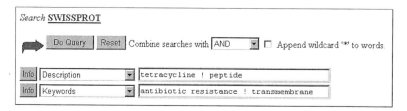

Figure 8 This query retrieves all entries on tetracycline that are relevant to antibiotic resistance but are neither trans-membrane proteins nor peptides. Note that one can directly search for the phrase 'antibiotic resistance' in the 'Keyword' index. To search for entries described by 'resistance protein'. one has to ask for each word separately (namely, 'resistance & protein') because the index for 'Description' only contains single words (cf. Index Browser, above).

and that expressions are evaluated strictly left to right (all operators have the same precedence). However, the drop-down menu 'Combine searches with' can be used to select a logical operator that will be applied to join the lines of the query form. *Figure 8* shows a query that retrieves all entries on tetracycline that are relevant to antibiotic resistance but are neither trans-membrane proteins nor peptides.

3.1.4 The Query Manager

Often it is helpful or even necessary to break complex operations into smaller, more manageable tasks. SRS stores user queries and makes them available for later reuse through the Query Manager. Selected query results may be annotated, saved to disk, deleted, displayed in particular Views, or combined using logical operators.

Given the previous example, consider the wish to exclude those entries with an uncertain sequence. Following the next few steps as outlined below will accomplish just that. To avoid editing the original query, which one might want

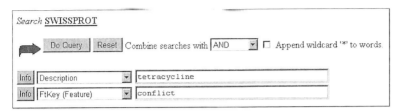

Figure 9 A query to retrieve entries described as 'tetracycline' and with a sequence about which there is disagreement in the available literature.

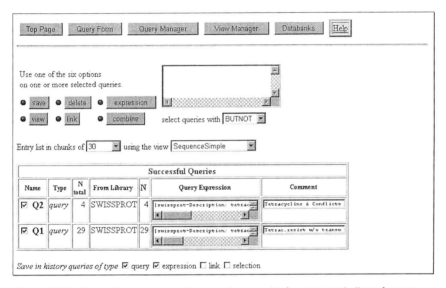

Figure 10 The Query Manager page offers previous results for reuse and allows for more complex operations.

to keep for reference, run a new query as shown in *Figure 9*. Go to the Query Manager page, select the last two entries, and finally choose 'BUTNOT' as operator and click on the 'combine' button (*Figure 10*).

Advanced users will learn to appreciate the option to enter more complex query expressions directly. The previous result can also be obtained by entering '[SWISSPROT-Description: tetracycline ! peptide] & [SWISSPROT-Keywords: antibiotic resistance ! transmembrane] ! ([SWISSPROT-FtKey: conflict] > parent)' (without the quotes) into the window next to the 'expression' button. Press the 'expression' button to request the evaluation of the query. This example demonstrates several features:

(a) Simple queries take the form '[DatabaseName–FieldName:QueryString]'. Logical operators may be used with the same restrictions as in the Query Form. For brevity, the abbreviation of the field name, as displayed in the Index Browser, can be used, e.g. '[SWISSPROT-des:tetracyc*]' to query the 'Description' field.

(b) Queries can be combined using logical operators. Here, parentheses *may* be used for grouping.

(c) The syntax for a link query is 'A < B' or 'A > B', where A and B can each be a database name or the name of a set of entries (such as those automatically assigned, e.g. 'Q2'). The first expression retrieves those entries of A that are linked to B, while the second one retrieves the entries of B that are linked to A (as suggested by the points of the 'arrows'). Thus, 'A > B' and 'B < A' are equivalent expressions.

(d) *Special links:* Some entries can have sub-entries (such as the SWISS-PROT Features). These are indexed and can be queried just like the main entries. Linking with the special pre-defined 'parent' provides a mechanism to access the entries that contain the retrieved sub-entries. There are other special links for moving up and down in hierarchical databases and to link hits of a search application to the originally searched database (. . .searchdb').

SRS Links can always be searched in both directions; they are bi-directional. If a specialized database explicitly refers to a common (e.g. repository) database, one can thus also request linked entries of the specialized database starting from an entry in the common database. The designers of the common database do not need to know even of the existence of the specialized database. SRS also finds entries that are linked via an intermediary. For example, a SWISS-PROT entry that has links to both EMBL and PDB creates (indirect) links between the respective EMBL and PDB entries. Consequentially, indirect links can involve any number of intermediate linking steps. SRS uses the shortest way through the net of databases when resolving a link. SRS server administrators can set up penalties reflecting the desirability of using a particular inter-database link. Users can chose a different path through the network by explicitly requesting a series of links (. . . A > B > ... > C').

3.2 Creating custom Views

The simplest way to have query results displayed in a custom View is specifying the fields of interest in the Query Form. There are two conceptually different View types that can be requested.

(a) **List Views** usually try to conserve as much as possible of the entry as formatted in the original database text file. List Views thus typically act as a filter, displaying only those lines of the entry that have information pertaining to the selected fields. Often, the text is pretty-printed and contains hyperlinks where appropriate. It is not, however, altered in content. List Views are best employed when information in a database entry needs to be read in its original context. This is typically the case for fields that implicitly or explicitly refer to information elsewhere in the entry. *Figure 11* shows an excerpt of an entry that would considerably lose in clarity when displayed out of context as fields in a table.

(b) **Table Views** give a better overview by collecting only requested data for display. Values are extracted from the original entry format of the database text file, and the data are usually re-formatted for easy readability and concise display. Table Views are ideal for summaries, excerpts, and also when data from different databases needs to be combined. All the Views shown in *Figure 3*, *Figure 4*, and *Figure 5* are Table Views.

The View Manager allows the creation of more complicated Views, such as the one show in *Figure 5b*. Also, Views can be deleted. Named custom Views can be created from scratch, or derived by editing existing Views. Named custom Views are then available just like the pre-defined Views. Users can specify the type of the View (List View or Table View), and whether the abbreviated forms or the complete names of the displayed database fields are to be shown in Table Views. Views are defined for a set of databases (the *root*-libraries). Views also may include fields that are only present in some of the chosen root-libraries. Many elaborate Views link to additional databases that are to be specified (the *leaves* of the View). When all the involved databases have been selected, the user chooses the respective database fields that should be shown, and the formats in which to display data objects (e.g. sequences, alignments) of particular fields.

Advanced options allow the following:

(a) 'Display only number of linked entries': The query that links a displayed entry to a leaf database may result in a set of several entries. For summaries, the size of this set (instead of the individual entries) can be displayed; this includes a hyper-link to the set of entries for inspection of details.

(b) 'Use view to display entries': In List Views, the entries of each leaf-database can be shown in a specified View.

```
┌ SWISSPROTTER1 ECOLI

AC   P03038;
DE   TETRACYCLINE REPRESSOR PROTEIN CLASS A (TRANSPOSON 1721).
RN   [1]
RA   ALLMEIER H., CRESNAR B., GRECK M., SCHMITT R.;
RL   GENE 111:11-20(1992).
RN   [2]
RA   TRUEMAN P., SHARPE G.S., BARTH P.T.;
RL   SUBMITTED (NOV-1993) TO EMBL/GENBANK/DDBJ DATA BANKS.
RN   [3]
RA   WATERS S.H., ROGOWSKY P., GRINSTED J., ALTENBUCHNER J., SCHMITT R.;
RL   NUCLEIC ACIDS RES. 11:6089-6105(1983).
FT   DNA_BIND    26    45       H-T-H MOTIF.
FT   SITE        64    64       INVOLVED IN BINDING TO [MG-TC]+ (BY
FT                              SIMILARITY).
FT   METAL       100   100      MAGNESIUM (OF [MG-TC]+ COMPLEX) (BY
FT                              SIMILARITY).
FT   CONFLICT    65    66       TH -> ST (IN REF. 3).
FT   CONFLICT    80    80       I -> T (IN REF. 3).
FT   CONFLICT    154   155      DA -> ES (IN REF. 3).
SQ   Sequence    216 AA;
     MTKLQPNTVI RAALDLLNEV GVDGLTTRKL AERLGVQQPA LYWHFRNKRA LLDALAEAML
     AENHTHSVPR ADDDURSFLI GNARSFRQAL LAYRDGARIH AGTRPGAPQM ETADAQLRFL
     CEAGFSAGDA VNALMTISYF TVGAVLEEQA GDSDAGERGG TVEQAPLSPL LRAAIDAFDE
     AGPDAAFEQG LAVIVDGLAK RRLVVRNVEG PRKGDD
//
```

Figure 11 For this List View, the fields 'AccNumber' (accession number), 'Description', 'CitationNo', 'Authors', 'Citation', 'Sequence', and 'FtDescription' (feature sub-entry description) have been selected for display. The 'swiss' (SWISS-PROT) format has been chosen for the sequence field.

231

(c) 'Use query instead of link': This allows complex operations, such as performing a link following a specified path through the net of databases, or excluding certain entries. The set 'entry' holds the entry to be displayed and can be used in the query expression.

3.3 SRS world wide: using DATABANKS

The public network of SRS servers currently provides the scientific community with access to ~350 different databases. The 40 sites located in 26 countries offer a total of ~1300 databank copies.

To facilitate identification of databases of interest and choosing an appropriate server, a new component of SRS version 5.1 generates DATABANKS, a database of databanks, by traversing the public SRS servers around the world (4). Automated nightly compilation of the data guarantees that the catalogue is up to date.

SRS provides a framework for database documentation in form of the 'databank information page', which has a standardized layout for easy reference and includes a general description, references and internet links, as well as detailed documentation of database fields. Each entry in DATABANKS contains a copy of the SRS databank information page as provided by the server it was collected from, and it concludes with an overview of alternate sites. A typical entry is shown in *Figure 12*. If a stable connection to a particular site could not be established, the site is moved to the end of the list of alternatives. In these cases, data from previous runs are used as backup. A record of when the backup was originally retrieved indicates whether it might be out of date.

DATABANKS is typically searched by databank name or description. Protocol 6 is a guide to solving common questions using DATABANKS. More generally, any field present in the databank information pages, as well as site and server characteristics, can be queried.

Protocol 6

Search SRS world wide

Query DATABANKS

1 Connect to an SRS server that offers DATABANKS, for example *http://srs.ebi.ac.uk/* at EMBL-EBI.

2 Select 'SRS World Wide' from the SRS home page. This leads to a list of known public SRS servers.

3 Select 'Search' to get to the Query Form for DATABANKS.

Identification of databases of interest

1 Enter a query into the form. (See *Figure 13* for an example.)

2 Restrict the search to a subset of DATABANKS that includes only one site from each group of alternatives[a] as shown in *Figure 13*.[b]

3 Press the 'Continue' button to submit your query.

Choosing an appropriate server

1 Go back to the DATABANKS Query Form.

2 Enter your query into the form, e.g. ask for databanks named 'PIRALN' (without the quotes).

3 Press the 'Continue' button to submit your query.

[a] Currently, this representative site is chosen as the site that has the most extensive databank information page.

[b] This requests the 'selection flag' by including '[DATABANKS-SelFlag:*]' in your query.

Consider a user who is interested in databases offering sequence alignments, which hold information on well-characterized protein domains or families and can be used for functional assignments or phylogenetic examinations. Selecting the field 'Description' in the query form and asking for 'sequence' and 'align', yields a list of approximately 60 databank copies. By restricting the search as instructed by Protocol 6, part B; however, the query only fetches a more manageable list of 15 representative databanks. In addition to general databases of protein domains or families such as PFAM (12), PRINTS (13), or PIRALN (14), a user will also find specialized databases, such as HOVERGEN (vertebrates, 15), AMmtDB (vertebrate mitochondria, 16), RDP (ribosomes, 17), FSSP and HSSP (protein structure, 18 and 19), or TRANSFAC (transcription factors, 20). The user can now browse the descriptions of the databases retrieved and refine or broaden the search.

Having decided on a particular database, e.g. PIRALN, one usually finds that more than one server maintains a copy of the database, and the list of results shows alternative sites (see *Figure 14*). The number of indexed entries and the release number (where assigned by the server maintainers) help users to choose a nearby site that has a current version of the database.

Both the list of results and the overview of alternate sites compiled for each entry of DATABANKS provide direct links for remote queries. These lead to the respective query forms of the databases at the remote sites as specified.

3.4 Interfacing with SRS over the network

This section demonstrates with a few examples how to access SRS directly over a network connection. Currently, that means using the Web to request specific parts of the SRS Web interface. The Web interface is rendered by the program 'wgetz', which sites may provide in different locations, e.g. 'http://srs.ebi. ac.uk/srs5bin/cgi-bin/wgetz' for EMBL-EBI. An easy way of getting the path to wgetz is looking at the 'Databanks' hyper-link on the SRS home page. To request the launch of wgetz, append the character '?' and any parameters that you wish to supply. If you specify more than one parameter, separate them

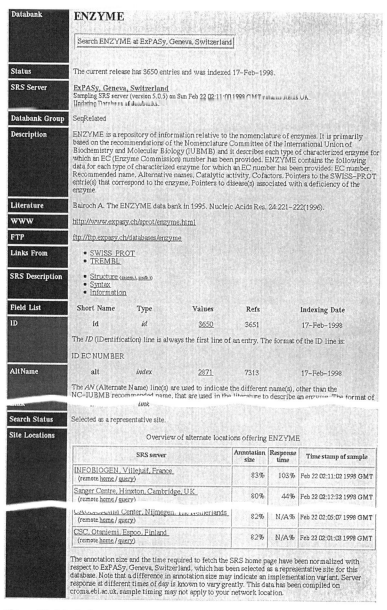

Figure 12 A typical DATABANKS entry. The entry contains a copy of the respective remote SRS databank information page, which includes a description, references and links, as well as detailed documentation of database fields and indices. It concludes with a listing of alternative sites that offer ENZYME. Direct links to these sites and the remote query forms for ENZYME are provided. For uses in the network vicinity of a particular DATABANKS server, the relative response times compiled by that server give a clue to the net distances to other sites ('N/A' indicates problems connecting at the specified time).

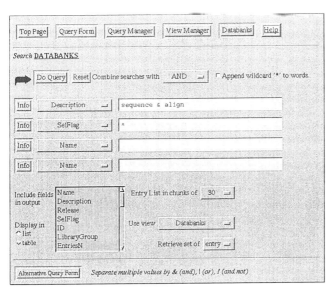

Figure 13 Query for databanks that have a description containing the terms 'sequence' and 'align'. The second line of the query form requests that the results be restricted to one representative databank for each group of alternatives.

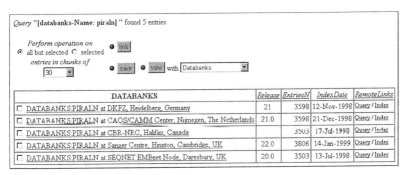

Figure 14 The results of a query for databanks named 'PIRALN'. The number of indexed entries and the release number (where assigned by the server maintainers) help users to choose a nearby server that offers a current version of the appropriate database.

with '+' instead of with spaces. To include literal spaces in your parameters, replace them with '%20'.

There are various ways in which entries can be displayed, for example:

(a) **As plain lists of entries**: Specify your query as you would in the expression box of the Query Manager.

(b) **As individual entries**: Specify the '–e' (entry) switch together with an SRS query, e.g. .../wgetz?-e+[SWISSPROT-ID:LPTR_BACST].

(c) **As lists of entries that can be browsed**: Specify the '-sl' (sequence list) switch with an SRS query. This yields a page that looks like the list of results in the standard SRS Web interface.

235

For each of the above, a particular View may be requested, e.g. '-view + SequenceSimple'.

Some SRS functions can be accessed directly as entry points into the system, e.g. requesting the query form for a specified set of databases. The switch '-l' (libraries) sets the databases to operate on, the '-fun' (function) switch selects the operation requested: e.g. '../wgetz?-fun+PagoQueryForm+-l+SWISS PROT%20SWISSNEW'.

Coming back to a previous session, or to maintain data across several wgetz calls, a user context needs to be specified with the '-id' switch. The id of a session is part of most of the links it displays; it is easily seen in the 'resume' link on the Top Page (cf. *Figure 6*). Chapter 3 of the on-line manual contains further information on linking to SRS servers using the Web.

Future versions of SRS will also support other mechanisms of remote access to SRS servers. We have already developed a prototype Corba server. Generally, SRS is able to serve structured data and methods that operate on them to any object oriented or procedural environment.

4 SRS server side

SRS may be installed locally on your site. This section outlines how to take advantage of a local SRS system.

4.1 User's point of view

In addition to accessing their SRS system over the Web, local users can run queries from the command line, use scripting to automate repetitive tasks or perform elaborate analyses, and modify parsers (if the administrators of the system permit that).

To activate a particular installation of SRS on your site, use '. prep_srs.sh' from Bourne, Korn or POSIX shells, or 'source prep_srs' from C shells. If the sourced files are not in your path, they are provided in the 'etc' sub-directory of the place where SRS was installed (the SRS *root* directory). After this initial-ization, one can use the 'getz' program to query the SRS system; e.g. 'getz [SWISSPROT-Des:tetracycline] -f des' will print those lines of the matching entries that hold the Description field. Use Chapter 4 of the on-line manual as a reference and for more examples.

While 'getz' is quite popular among users and is frequently used within shell scripts or Perl programs, it is inefficient for repeated fine-grained access to data. The only way to avoid a huge number of 'getz' invocations is to request many fields in one query, and users often end up parsing the resulting 'getz' output—what a waste considering that SRS has already parsed the entry into fields! A better solution is integrating a programming language with SRS:

(a) The entire SRS functionality can be accessed through the C application pro-gramming interface (API). See Chapter 12 of the on-line manual for instruc-tions on linking your C code with the SRS library, and several examples.

More examples can be found in the 'demo' sub-directory of the SRS installation. This is the interface to use for the most extensive access to SRS features. However, only the functions introduced in the examples are guaranteed to be supported in future versions—other functions may change as we improve and extend the system.

(b) More and more, Icarus allows access to SRS functionality. Already now, SRS queries and access to token tables is supported. Token tables hold the results of the parsing process. In future versions, application launching and all other SRS functions and data will be accessible through Icarus. The example shown in *Figure 15a* prints the entry-ids and the molecular weights of all matches to the query 'SWISSPROT-des:Tetracycline*'. The example shown in *Figure 15b* takes an SRS query string on the command line and dumps the 'fields' token table. See Chapters 5-7 of the on-line manual to learn more about Icarus.

(c) Future versions of SRS will offer native language interfaces to other general purpose languages like C++ or Java, and to popular scripting languages like Perl (for which a prototype has already been developed)

Sometimes users find they struggle to extract particular data through a query. Often a simple modification or extension of the parsers involved helps a lot. Clearly, many users do not want to deal with the parsers themselves, or they even may not have permission to do so. Still, they need to know when it pays to ask for help. Typical problems that can be solved include:

```
#!/bin/env icarus

$my_set=$Query:'SWISSPROT-des:!etracycline*'
if:$my_set
{ foreach:[$entry in:$my_set]
  { $Print:| - Entry ($entry.name)
    foreach:[$token in:$entry.tokens:molweight]
      $Print:|  molecular weight = ($token.str)
    $entry.delete # free space allocated for $Entry object
  }
  $my_set.delete  # free space allocated for $Set object
}
$dp:''  # to force stdout flush
```

```
#!/bin/env icarus

$dp:''
if:$ArgN<2
{ $Print:||| usage:  ($Arg:1) {SRS-query}
  $dp:''  # to force stdout flush
  $Exit
}

$query=$Arg:2

$my_set=$Query:$query
if:$my_set
{ $Print:||| Set returned by query '$query'
  foreach:[$entry in:$my_set]
  { $Print:||| Entry ($entry.name)
    foreach:[$token in:$entry.tokens:fields]
    { $Print:" - token code: "  $Print:$token.code
      $Print:"  token string:\n" $Print:$token.str
    }
    $entry.delete # free space allocated for $Entry object
  }
  $my_set.delete  # free space allocated for $Set object
}

$dp:''  # to force stdout flush
```

Figure 15 Example for using Icarus as a cripting language (for SRS version 5.1 or higher). The program displayed on the left prints the entry-ids and the molecular weights of all matches to the query 'SWISSPROT-des:Tetracycline*'. The program shown on the right takes an SRS query string as its argument on the command line and prints the parser's 'fields' token table: For each token, it shows both the *token code* (a label, which can be used to determine how a token is further processed), and the *token string*. The *token object* (which is used, e.g. to hold the sequence object) could have been accessed using $token.obj. $Query returns a $Set object. See Chapter 7 of the on-line manual for a reference of Icarus classes and chapters 5 and 6 to learn more about Icarus syntax and functions.

(a) **Extracting data from a field**: If you find yourself trying to query certain *parts* of a field regularly, that extraction process should be delegated to the parser. In a parser, this extraction can also be much more sophisticated.

(b) **Querying phrases**: The parser controls which terms can be matched by a query (see Sections 3.1.1 and 3.1.2). Sometimes the way a parser breaks a field into indexed terms is not well suited for answering a particular question. Changing the parser or adding a new field that takes particular requirements into account solves the problem. Consider a comment field, for which the parser extracts single words for indexing; the query 'not & known' will not only find the phrase 'not known', but also retrieve instances that contain the two words out of context. To address this problem, the parser might be changed to watch, e. g., for words preceded by 'not', and write the two word phrase to the index.

(c) Similarly, queries that use logical operators to combine data from several fields are sometimes not sufficient if the necessary context has been lost in the indexing process. Changing the parser is again the only satisfactory solution.

4.2 Administrator's point of view

Within the scope of this chapter, we can only give some help in getting an SRS system up and running. Each of the following topics, however, should be of further interest to SRS server administrators:

(a) The automation of tasks in server, database, and application maintenance.

(b) The design and implementation of new modules for the integration of databases and application programs. This must include a discussion of lazy, forced, and explicit parsing, and how to best make use of this layered approach.

(c) A manual of Icarus that introduces new users, and serves as a reference to experienced programmers. Icarus is an efficient, expressive scripting language that has been particularly designed for the purpose of describing data-structures and the rapid development of highly flexible parsers. Yet Icarus has evolved towards a general-purpose object-oriented programming language, which we expect to play a central role in scripting and large-scale data analysis in the future of SRS.

Clearly, these issues are all rather complex and deserve a dedicated discussion in themselves.

4.2.1 How to get your own SRS system

Download SRS from *ftp://ftp.ebi.ac.uk/pub/software/unix/srs/*, the current version is in file 'srs5.1.tar.gz' and is freely available to the public. Please note that future versions of SRS will continue to be freely available for academic non-commercial use. Future versions of SRS and services are provided by LION Bioscience Ltd (*http://www.lionbio.co.uk/*).

To install SRS in a place of your choice, unpack the distribution file (e.g. 'gzip —cd srs5.1.tar.gz | tar —xf -'). This creates a directory for SRS, which we call the SRS *root directory*, and which we will print as '...' for the rest of this chapter. From this directory, first run './srsinstall all'. To also install the SRS web interface, run './srsinstall www' next. At completion, this prints two lines that have to be inserted into the 'srm.conf' file of your web server. Ask your system administrator to do this if necessary. If there is no web server installed on your site, save the lines for later reference. You then need to have a Web server installed. We now normally use Apache (see *http://www.apache.org/*).

4.2.2 Basic configuration steps

SRS obviously comes without any databases or applications. However, a large number of *database modules* come with the distribution. These modules allow SRS to index and access the respective databases. The files that constitute a module are stored in '.../icarus/db/'. For each database, respectively, there is:

(a) A '*module-name*.is'-file that holds the parser.

(b) A '*module-name*.i'-file defines the database fields and links that can be queried.

(c) A '*module-name*.it'-file that contains optional database documentation, which is used to construct the databank information page.

The databank information pages usually contain fields that report database sources, indicating where the required files can be downloaded. This can directly be read in the '*module-name*.it'-files. More conveniently, this information can be queried for in the Database of Databanks as introduced in Section 3.3. This also gives access to the hundreds of modules developed worldwide, of which only some are included with the standard distribution. If a database module is offered by remote servers only, download the files that constitute the respective database module by following the hyper-links provided in the 'SRS Description' field of the DATABANKS entry.

Edit '.../icarus/db/srsdb.i' to reflect the databases available at your installation:

(a) There needs to be a 'file:*module-name*.i' command to include the appropriate field definitions.

(b) Each database needs a unique id number: Edit the list 'libIds' accordingly. When adding a new database, specify the database as defined by '$Library' in the '*module-name*.i'-file.

(c) The locations of the database files are declared in the 'libs' list of '$LibLoc' statements. SRS uses this list to determine which databases are available. Therefore, remove statements for databases not kept at your site, or turn them into comment lines, which start with a hash ('#'). Edit the other directories as appropriate. Again, when adding a new database, specify the database as defined by $Library in the '*module-name*.i'-file.

To commit your changes, run 'srssection'. In preparation for this or any other SRS program, initialize your environment: Users of Bourne, Korn or POSIX shells execute '`..../etc/prep_srs.sh`', while C shell users '`source .../etc/ prep_srs`'. (This sets the path for executables, the environment variable '`$SRSROOT`' and several others.)

All of the above also applies to modules for application programs, only they are not yet included in DATABANKS.

4.2.3 Prepare your databases for access through SRS

After preparing the appropriate database text files and configuring SRS accordingly, a set of indices has to be created for each database. These indices allow fast queries of database fields and links.

The program '`srscheck`' generates the script '`srsupdate`' and shows which indices need to be created or updated. Use the '`-l`' (library) switch to check the indices of a particular database only. Run `srsupdate` to actually build the required indices. For small databases, this is a matter of minutes. Indexing large databases like EMBL may take several hours, though. At successful completion, your databases are ready for SRS queries.

5 Where to turn to for help

If you experience difficulties, do not despair!

(a) If you have installation problems, please read carefully through the README file that comes with the distribution.

(b) Try to make good use of the on-line manuals. While it is sometimes difficult to find particular bits of information there, they are quite extensive, and should certainly be consulted first. As the individual sections can be quite large, using the text search feature of your browser can be quite helpful. Users will want to focus on Chapters 1 and 2, which deal with the Web interface and the Query language, respectively. Administrators should start with Chapters 4, 10, and 11. Chapters 10 and 11 contain a lot of information on installation and set-up, and Chapter 4 is a reference of SRS programs available from the command line. Use the to search within chapters, and do not be fooled by the section headings. Please note that the chapter numbers refer to the *on-line manual* of *SRS-5.1*.

(c) Try asking colleagues for help. There is a strong community of SRS aficionados out there that will be happy to help (that is, *if* you have looked through the on-line manual first).

(d) Send your question to *news://localnews/bionet.software.srs*, a newsgroup dedicated to SRS related questions.

(e) If you discover a bug, or have a problem of very technical nature, report it by e-mail to *srsdev@ebi.ac.uk*.

If you need more help configuring and customising your system, you may want to learn about SRS workshops and consultancy available from LION Bioscience Ltd (*srs-info@lionbio.co.uk*).

Acknowledgements

Many people have contributed to the wealth of SRS database and application modules that is publicly available now, and we are indebted to them all!

We wish to warmly thank Rob Falla for his help and the fruitful late-night discussions. Also, we gratefully thank Mark Wooding who thoroughly examined the final draft of this chapter. Any errors were certainly introduced afterwards!

References

1. Discala, C., *et al.* (1998). DBCAT, the public catalog of databases. INFOBIOGEN, Villejuif, France;contact `discala@infobiogen.fr`.
2. Frishman, D., Heumann, K., Lesk, A., and Mewes, H.-W. (1998). *Bioinformatics*, **14**, 551.
3. Brenner, S. E. (1995). *Science*, **268**, 622.
4. Kreil, D. P. and Etzold, T. (1998). *Trends Biochem. Sci.*, **24**, 155.
5. Etzold, T., Ulyanov, A., and Argos, P. (1996). In *Methods in enzymology* (ed. R. F. Doolittle). Vol. 266, p. 114. Academic Press.
6. Etzold, T. and Argos, P. (1993). *Comput. Appl. Biosci.*, **9**, 49.
7. Etzold, T. and Argos, P. (1993). *Comput. Appl. Biosci.*, **9**, 59.
8. Markowitz, V. M. and Ritter, O. (1995). *J. Comput. Biol.*, **2**, 537.
9. Davidson, S. B., Overton, C., Tannen, V., and Wong, L. (1997). *Int. J. Digit. Libr.*, **1**, 36.
10. Altschul, S. F., *et al.* (1990). *J. Mol. Biol.*, **215**, 403.
11. Thompson, J. D., Higgins, D. G., and Gibson, T .J. (1994). *Nucleic Acids Res.*, **22**, 4673.
12. Sonnhammer, E. L., *et al.* (1998). *Nucleic Acids Res.*, **26**, 320.
13. Attwood, T. K., Beck, M. E., Bleasby, A. J., and Parry-Smith, D. J. (1994). *Nucleic Acids Res.*, **24**, 182.
14. Barker, W. C., *et al.* (1998). *Nucleic Acids Res.*, **26**, 27.
15. Duret, L., Mouchiroud, D., and Gouy, M. (1994). *Nucleic Acids Res.*, **22**, 2360.
16. Lanave, C., *et al.* (1999). *Nucleic Acids Res.*, **27**, 134.
17. Maidak, B. L., *et al.* (1997). *Nucleic Acids Res.*, **25**, 109.
18. Holm, L., *et al.* (1992). *Protein Sci.*, **1**, 1691.
19. Sander, C. and Schneider, R. (1991). *Proteins*, **9**, 56.
20. Heinemeyer, T., *et al.* (1998). *Nucleic Acids Res.*, **26**, 362.

List of suppliers

Anderman and Co. Ltd., 145 London Road, Kingston-upon-Thames, Surrey KT2 6NH, UK.
Tel: 0181 541 0035   Fax: 0181 541 0623

Beckman Coulter (UK) Ltd., Oakley Court, Kingsmead Business Park, London Road, High Wycombe, Buckinghamshire HP11 1JU, UK.
Tel: 01494 441181
Fax: 01494 447558
URL: http://www.beckman.com
Beckman Coulter Inc., 4300 N Harbor Boulevard, PO Box 3100, Fullerton, CA 92834-3100, USA.
Tel: 001 714 871 4848
Fax: 001 714 773 8283
URL: http://www.beckman.com

Becton Dickinson and Co., 21 Between Towns Road, Cowley, Oxford OX4 3LY, UK.
Tel: 01865 748844
Fax: 01865 781627
URL: http://www.bd.com
Becton Dickinson and Co., 1 Becton Drive, Franklin Lakes, NJ 07417-1883, USA.
Tel: 001 201 847 6800
URL: http://www.bd.com

Bio 101 Inc., c/o Anachem Ltd., Anachem House, 20 Charles Street, Luton, Bedfordshire LU2 0EB, UK.
Tel: 01582 456666   Fax: 01582 391768
URL: http://www.anachem.co.uk

Bio 101 Inc., PO Box 2284, La Jolla, CA 92038-2284, USA.   Tel: 001 760 598 7299
Fax: 001 760 598 0116
URL: http://www.bio101.com

Bio-Rad Laboratories Ltd., Bio-Rad House, Maylands Avenue, Hemel Hempstead, Hertfordshire HP2 7TD, UK.
Tel: 0181 328 2000   Fax: 0181 328 2550
URL: http://www.bio-rad.com
Bio-Rad Laboratories Ltd., Division Headquarters, 1000 Alfred Noble Drive, Hercules, CA 94547, USA.
Tel: 001 510 724 7000
Fax: 001 510 741 5817
URL: http://www.bio-rad.com

CP Instrument Co. Ltd., PO Box 22, Bishop Stortford, Hertfordshire CM23 3DX, UK.
Tel: 01279 757711   Fax: 01279 755785
URL: http://www.cpinstrument.co.uk

Dupont (UK) Ltd., Industrial Products Division, Wedgwood Way, Stevenage, Hertfordshire SG1 4QN, UK.
Tel: 01438 734000
Fax: 01438 734382
URL: http://www.dupont.com
Dupont Co. (Biotechnology Systems Division), PO Box 80024, Wilmington, DE 19880-002, USA.
Tel: 001 302 774 1000
Fax: 001 302 774 7321
URL: http://www.dupont.com

Eastman Chemical Co., 100 North Eastman Road, PO Box 511, Kingsport, TN 37662-5075, USA.
Tel: 001 423 229 2000
URL: http://www.eastman.com

Fisher Scientific UK Ltd., Bishop Meadow Road, Loughborough, Leicestershire LE11 5RG, UK.
Tel: 01509 231166
Fax: 01509 231893
URL: http://www.fisher.co.uk
Fisher Scientific, Fisher Research, 2761 Walnut Avenue, Tustin, CA 92780, USA.
Tel: 001 714 669 4600
Fax: 001 714 669 1613
URL: http://www.fishersci.com

Fluka, PO Box 2060, Milwaukee, WI 53201, USA.
Tel: 001 414 273 5013
Fax: 001 414 2734979
URL: http://www.sigma-aldrich.com
Fluka Chemical Co. Ltd., PO Box 260, CH-9471, Buchs, Switzerland.
Tel: 0041 81 745 2828
Fax: 0041 81 756 5449
URL: http://www.sigma-aldrich.com

Hybaid Ltd., Action Court, Ashford Road, Ashford, Middlesex TW15 1XB, UK.
Tel: 01784 425000
Fax: 01784 248085
URL: http://www.hybaid.com
Hybaid US, 8 East Forge Parkway, Franklin, MA 02038, USA.
Tel: 001 508 541 6918
Fax: 001 508 541 3041
URL: http://www.hybaid.com

HyClone Laboratories, 1725 South HyClone Road, Logan, UT 84321, USA.
Tel: 001 435 753 4584
Fax: 001 435 753 4589
URL: http://www.hyclone.com

Invitrogen Corp., 1600 Faraday Avenue, Carlsbad, CA 92008, USA.
Tel: 001 760 603 7200
Fax: 001 760 603 7201
URL: http://www.invitrogen.com
Invitrogen BV, PO Box 2312, 9704 CH Groningen, The Netherlands.
Tel: 00800 5345 5345
Fax: 00800 7890 7890
URL: http://www.invitrogen.com

Life Technologies Ltd., PO Box 35, Free Fountain Drive, Incsinnan Business Park, Paisley PA4 9RF, UK.
Tel: 0800 269210
Fax: 0800 838380
URL: http://www.lifetech.com
Life Technologies Inc., 9800 Medical Center Drive, Rockville, MD 20850, USA.
Tel: 001 301 610 8000
URL: http://www.lifetech.com

Sharp & Dohme, Research Laboratories, Neuroscience Research Centre, Terlings Park, Harlow, Essex CM20 2QR, UK.
URL: http://www.msd-nrc.co.uk
MSD Sharp and Dohme GmbH, Lindenplatz 1, D-85540, Haar, Germany.
URL: http://www.msd-deutschland.com

Millipore (UK) Ltd., The Boulevard, Blackmoor Lane, Watford, Hertfordshire WD1 8YW, UK.
Tel: 01923 816375
Fax: 01923 818297
URL: http://www.millipore.com/local/UK.htm
Millipore Corp., 80 Ashby Road, Bedford, MA 01730, USA.
Tel: 001 800 645 5476
Fax: 001 800 645 5439
URL: http://www.millipore.com

New England Biolabs, 32 Tozer Road, Beverley, MA 01915-5510, USA.
Tel: 001 978 927 5054

Nikon Inc., 1300 Walt Whitman Road, Melville, NY 11747-3064, USA.
Tel: 001 516 547 4200
Fax: 001 516 547 0299
URL: http://www.nikonusa.com
Nikon Corp., Fuji Building, 2-3, 3-chome, Marunouchi, Chiyoda-ku, Tokyo 100, Japan.
Tel: 00813 3214 5311
Fax: 00813 3201 5856
URL: http://www.nikon.co.jp/main/index_e.htm

Nycomed Amersham plc, Amersham Place, Little Chalfont, Buckinghamshire HP7 9NA, UK.
Tel: 01494 544000   Fax: 01494 542266
URL: http://www.amersham.co.uk
Nycomed Amersham, 101 Carnegie Center, Princeton, NJ 08540, USA.
Tel: 001 609 514 6000
URL: http://www.amersham.co.uk

Perkin Elmer Ltd., Post Office Lane, Beaconsfield, Buckinghamshire HP9 1QA, UK.
Tel: 01494 676161
URL: http://www.perkin-elmer.com

Pharmacia Biotech (Biochrom) Ltd., Unit 22, Cambridge Science Park, Milton Road, Cambridge CB4 0FJ, UK.
Tel: 01223 423723
Fax: 01223 420164
URL: http://www.biochrom.co.uk
Pharmacia and Upjohn Ltd., Davy Avenue, Knowlhill, Milton Keynes, Buckinghamshire MK5 8PH, UK.
Tel: 01908 661101
Fax: 01908 690091
URL: http://www.eu.pnu.com

Promega UK Ltd., Delta House, Chilworth Research Centre, Southampton SO16 7NS, UK.
Tel: 0800 378994
Fax: 0800 181037
URL: http://www.promega.com

Promega Corp., 2800 Woods Hollow Road, Madison, WI 53711-5399, USA.
Tel: 001 608 274 4330
Fax: 001 608 277 2516
URL: http://www.promega.com

Qiagen UK Ltd., Boundary Court, Gatwick Road, Crawley, West Sussex RH10 2AX, UK.
Tel: 01293 422911   Fax: 01293 422922
URL: http://www.qiagen.com
Qiagen Inc., 28159 Avenue Stanford, Valencia, CA 91355, USA.
Tel: 001 800 426 8157
Fax: 001 800 718 2056
URL: http://www.qiagen.com

Roche Diagnostics Ltd., Bell Lane, Lewes, East Sussex BN7 1LG, UK.
Tel: 01273 484644   Fax: 01273 480266
URL: http://www.roche.com
Roche Diagnostics Corp., 9115 Hague Road, PO Box 50457, Indianapolis, IN 46256, USA.
Tel: 001 317 845 2358
Fax: 001 317 576 2126
URL: http://www.roche.com
Roche Diagnostics GmbH, Sandhoferstrasse 116, 68305 Mannheim, Germany.
Tel: 0049 621 759 4747
Fax: 0049 621 759 4002
URL: http://www.roche.com

Schleicher and Schuell Inc., Keene, NH 03431A, USA.
Tel: 001 603 357 2398

Shandon Scientific Ltd., 93-96 Chadwick Road, Astmoor, Runcorn, Cheshire WA7 1PR, UK.
Tel: 01928 566611
URL: http://www.shandon.com

Sigma-Aldrich Co. Ltd., The Old Brickyard, New Road, Gillingham, Dorset XP8 4XT, UK.
Tel: 01747 822211
Fax: 01747 823779
URL: http://www.sigma-aldrich.com

Sigma-Aldrich Co. Ltd., Fancy Road, Poole, Dorset BH12 4QH, UK.
Tel: 01202 722114
Fax: 01202 715460
URL: http://www.sigma-aldrich.com
Sigma Chemical Co., PO Box 14508, St Louis, MO 63178, USA.
Tel: 001 314 771 5765
Fax: 001 314 771 5757
URL: http://www.sigma-aldrich.com

Stratagene Inc., 11011 North Torrey Pines Road, La Jolla, CA 92037, USA.
Tel: 001 858 535 5400
URL: http://www.stratagene.com

Stratagene Europe, Gebouw California, Hogehilweg 15, 1101 CB Amsterdam Zuidoost, The Netherlands.
Tel: 00800 9100 9100
URL: http://www.stratagene.com

United States Biochemical, PO Box 22400, Cleveland, OH 44122, USA.
Tel: 001 216 464 9277

Index

Brackets denotes figures

1D-3D profile 5

alignment statistics
  background distribution 175
  significance 173

Bankit sequence submission
  program 209, [209]
Bayesian methods in HMMs
  87-8
BioCatalogue (web site) 192
BiowURLd (web site) 191
BLAST sequence search
  program 172-3, 185-7,
  [186], 199-200, 207, 223
BLOCKS sequence alignment
  database 81, 106, 147, 152
BLOSUM substitution matrix
  55, 183-4

CASP experiment 9
clique detection in structure
  comparison 28
CLUSTAL multiple sequence
  alignment program 61,
  [62], 207, 223
COBBLER sequence search
  program 108
codon based alignment 68
coiled-coil structure prediction
  137
comparative genomics 51

Compugen computer company
  201
consensus sequence 145, [146]

DALI structure comparison
  program 39
DATABANKS sequence retrieval
  program 232-3, [234], [235]
DBCAT database catalogue 216
DDBJ (web site) 195, 208
DEJAVU structure comparison
  program 40
deterministic pattern 146
DIALIGN multiple sequence
  alignment program 64
Dirchlet mixtures 84, 90, 147
DOMO sequence alignment
  database 106
double dynamic programming
  method 27, 39-40
DSC structure prediction
  method 127
DSSP secondray structure
  program 119
dynamic programming in
  structure comparison 20,
  26-8
dynamic programming in
  threading 5, 6

EBI (web site) 195, 205
EMBL sequence database 195,
  220
EMBnet 195-7
Entrez sequence retrieval
  program 203, 205, [206]

expectation value (e-value) 177,
  187
extreme value distribution
  175-6, [176]

FASTA sequence search
  program 173, 185-7, [186],
  201, 207
fold recognition 3, [4]

gap-BLAST sequence search
  program 107, 109
GA_FIT structure comparison
  program 36
genetic algorithm in structure
  comparison 36
GENFIT structure comparison
  program 42-6
GenTHREADER 8
global sequence alignment
  169-70
GOR structure prediction
  method 121
graph-matching in structure
  comparison 26, 28

hidden Markov models (HMM)
  77, 144
HMMER2 sequence alignment
  program 79, 82-4

Internet 191
iterated sequence search
  methods 99-100, 109

Jackknife test 119
JPRED structure prediction
    method 129

LAMA multiple sequence
    alignment program 110
local sequence alignment
    170-171
low-complexity sequence
    regions 53

MACAW multiple sequence
    alignment program 65, 98,
    101
Maestro sequence retrieval
    program 203
MAST sequence search program
    108
minimum description length
    148
motif 143
multiple alignment
    databases 72
    editing 69-72
    of sequences 57-60, 95-6
    profile 77-9, 86-7, 101-2,
    146-7, 185
multiple structure comparison
    41

NCBI (web site) 192, 195, 199
NEE-HOW project 191
neural network method 123
NNSSP structure prediction
    method 125

PAM (Dayhoff) substitution
    matrix 55, 182-3
Paracel computer company
    202, [205]
patten
    databases 150
    classification 149
    discovery 154, [156]
    sensitivity 150
    specificity 150
Pedro's BioMolecular Research
    Tools (web site) 191
Pfam protein family database
    80, 105, 153
PHD structure prediction
    program 7, 123

phylogenetic trees 98
position-specific scoring matrix
    (PSSM) 101, 185
PRALIGN multiple sequence
    alignment program 65
Pratt sequence pattern
    discovery program
    156-162, [157]
Pred2ary structure prediction
    program 124
PREDATOR structure prediction
    method 126-7
PRINTS sequence alignment
    database 81, 105, 147,
    152
probabilistic pattern 146
probability value (p-value) 177
    176
PROCLASS sequence alignment
    database 106-7
ProDom sequence alignment
    database 106
profile-HMMs 86-8, 146-7, 1
    88
progressive multiple alignment
    61-3
Prosite
    database 105, 145-7, 151-2,
    [152]
    sequence pattern 97, 144-6
    sequence profiles 80
protein comparison
    features 19
    relationships 20
protein family databases 105
protein secondary structure
    113-4, 116
    prediction 113
    from multiple alignment
    131-2, [133]
protein sequence
    databases 105, 179-81, 215
    Logo 72, 97, [97]
    motif 93, 143
    pattern 143
    pseudocounts 103
protein structure
    comparison of 15
    patterns in 143, 162-4
    topology [115], 116
psi-BLAST sequence search
    program 81-2, 107-9, 188

regular expression 146
residue pair potentials 6

root mean square deviation
    (RMSD) 26, 29-30, 163

Sakura sequence submission
    program 212, [212]
SARF2 structure comparison
    program 42-5
Seqin sequence submission
    program 209, [210]
sequence alignment 51, 168,
    [171]
    algorithm 169-172, [171],
    181
    gap penalty 56, 169-70,
    184
    global/local 54, [54]
    matrices 54-5, 182-4
sequence databank searching
    56-7, 103, 108, 167, 178
sequence homology 53
sequence motif 143
sequence retrieval 195, 203,
    [204]
SMART sequence alignment
    program 81
SPratt structure pattern
    discovery program 162-4
SRS (Sequence Retrieval System)
    195, 205, [207], 215, [218]
SSAP (SAP) structure comparison
    program 40, 164
SSEARCH sequence search
    program 185-7, [186]
SSPRED structure prediction
    method 129
structure comparison
    accuracy 29
    databases 46
structure environment 5, 6
structure prediction accuracy
    118
structure superposition 17
    multiple 21
    of globins 24
    using quaternions 23
superfold 2
SWISS-PROT 220-3, [222]

threading 1, 5, 138
    accuracy 8-9
    genome data 7, [8]
    protein domains 11
Time Logic computer company
    201, [202]

topological equivalence 25
TOPS topology program 116-7
transmembrane prediction of
    structure 133-6
        topology 136

VERTAA structure comparison
    program 37, [38]

Webin sequence submission
    program 210, [211]

World Wide Web (WWW) 191

z-score 178